THE
SHADOW
LADY

SHAWN ROBERTSON

Print ISBN: 978-1-09839-497-4
eBook ISBN: 978-1-09839-498-1

Printed in the United States of America

PROLOGUE

The Prophetic Dream

Hearing trembling, growling, groaning, stomping, and pounding. The electrical sparks illuminated so bright to the human, mortal eye, it seemed like in an instant the whole universe completely lit up. Silence filled the air no more. Then suddenly legions upon legions of immortals appeared out of nowhere, fighting each other to the death to defend the fate of humanity and the future of Mt. Olympus. Hera leads a revolt against the mighty Zues' to rewrite humanity in her own image. Thunderous sounds of iron swords clashing and exploding in the hands of GODS, GODDESSES, and TITANS, battling for the right to rule Mt. OLYMPUS. Hera says to the mighty Zues, "Your time of ruling is over, no more of your contradicting promises of hollowed-out truth." Zues responds, "You insolent, disrespectful, undeserving of the title, Queen of the Gods, you treacherous beast, never will I give in to your demand." They fight on and on seemed like centuries upon centuries. Until finally, Zues, Hera, and their mother, Rhea, were the only ones standing.

They fight and fight on and on til' both were utterly and completely exhausted, as Hera and Zues purposely but only for a brief moment separate from each other. Knowing their fate was closely staring them in the face, they use the last of their power to create their prophecy, to stop the other from gaining complete conquest, in the next life.

Then Zues without warning threw everything he had toward Hera! In that moment, without thought, Rhea stepped into the blast absorbing most of the wrath of Zues' final assault!

Rhea began to fade away into oblivion but not before she shields her daughter away inside an eternity stone with hopes of saving her for another time. Then Rhea looks back at Zues for the last time as she hurls the eternity stone into the blackness of space; she screams out to Zues saying, "You lose!"

Zues replies, "What have you done, you fool!" Before he could finish, there was a high-pitched sound so loud, it resembled a clock alarm, then everything fades to black.

CHAPTER 1

Morning News

6:30 in the morning on a Monday morning the alarm goes off. Elise slams her hand down hard on the clock that wouldn't go off. It just kept ringing and ringing. She snatched the clock off of her nightstand. Throwing it onto the floor, she comes to her senses saying out loud, "Damn, what kind of dream was that! I really need to stop watching a lot of SCI-FI movies before bed; that was some fucked-up shit!" She gradually gets up, slowly.

The same routine day in and day out. Elise fixes herself a hot cup of coffee. She grabs two pieces of bread out of the cupboard and places them in a toaster, then goes to the refrigerator to get some butter.

As Elise opens the door to the refrigerator, she notices a message on the door of the freezer. It was a note from her mother. It read: "Hello, baby girl. I know your job keeps you so busy. So, I took the liberty of washing your dirty clothes. I fixed your lunches for the week. I know, I know, I did not have to, but that's what moms are for. I would really be glad when you find someone to settle down with. LOL. OK, I will stop for now. Love always, your mom." As Elise finishes reading the note left on the refrigerator by her mom, she has a very loving smile on her face.

Thinking to herself, what am I going to do about that woman. She is always bugging me about a man. "If it is meant to be, it will happen." She speaks out loud to herself as if she were speaking with her mother. "I love you too, Mom, always." Her mother and she had a gesture of crossing their hearts and kissing their two fingers in the form of a peace sign, then throwing their love to each other.

Elise only has one sibling, a younger sister, who really looks up to her big sister. Her name is Kawanna Bishop.

Kawanna stands about five feet, four inches and 120 pounds, with a smooth complexion of caramel colored skin, but what made her so perfect was her long bouncing sandy reddish-black hair she wore mostly in a beautiful afro. She's also very athletic with a tight, lean body, like a Zumba instructor, a very health conscious person. Kawanna is a very headstrong young lady that always speaks her mind. She has a very sophisticated personality, always keeping up with the latest fads and trends. She loves art, poetry, and music and is a current sophomore in college. Her major is English literature.

She has her little circle of friends. They are just as passionate about art, poetry, and music. One of the three, in her little circle of friends, his full name is Chris Kencade, a junior majoring in political science.

Rex Singleton is number two, a diamond in the rough of Puerto Rican descent. He has that gothic thing going on, thinking the world is out to get him. There's nothing else to say about that; you know the rest.

Last but not least number three, we have Cindy Chase. Cindy is a very nerdy, red-headed, freckle-faced girl. She's an exchange student from a very small village out of Ireland. She is truly a mind that's ahead of her time. Cindy knows a lot of things particularly dealing with computers and physics equations relating to science.

This is the whole group who calls themselves *the Mad Pack*. One can only imagine what fate on earth could have brought these completely different individuals in the path of becoming associates, not to mention the best of friends.

No matter what is going on, they all make "chill" time for each other, at least twice a week. They always meet up at Lucy's Sports Bar and Grill.

They know the owner; his nickname is Lucky. Kawanna and Elise's mom went to school with the owner's wife. Her name is Jessie Ann Crawford. An old girlfriend from way back in the day.

Elise snaps out of her thoughts. Then hears movement outside her apartment door.

CHAPTER 2

The Package

There was a knock at Elise's door. She quickly places her toast on a plate as she hurries for the door. "Who is it?" Elise says as she gathers herself together. Concentrating on putting on her housecoat, while at the same time running her fingers thru her hair. Thinking to herself and wondering who the hell is at her front door this early in the morning.

She hears another knock at the door. She speaks in a loud and more aggressive voice, saying to the person on the other side of the door, "I SAID HOLD ON, JUST A SECOND. I'm COMING!"

A voice on the other side of the door finally responds replying, "Ok. I am sorry, ma'am."

Elise makes her way to the door in a rush; she opens the door. There standing in the doorway was a tall skinny white man, wearing black-and-white biker short speedos, a white helmet. On his face he was wearing black square lens glasses, resembling a Harry Potter and Peter Parker combination, a look-alike if you can imagine that.

He reaches inside of his beat-up western antique brown leather satchel, resembling a long and forgotten past, most likely an artifact given as a gift from a highly respected elderly family member. He pulls out a package to hand to Elise saying, "I have a package for Elise Bishop." The messenger hands her his clipboard and asks, "Can you please sign here and initial here?" Then he hands her a small box-shaped package about the size of a small DVD player.

She tips the messenger, thanks him, and sends him on his way. Elise shuts the door. But before she could walk away from the door, there is another knock at the door.

Elise yells out, "WHO IS IT NOW?"

A soft-spoken male voice replies, "It's Sherif, your landlord, ma'am."

She opens the door trying to hide her frustration. There standing in the doorway was a five feet, seven inches, 140 pounds Indian man with very dark skin like bronze, wearing silver rimmed glasses, shaggy curly hair, button-down shirt, blue jeans, and some worn-out leather sandals, reeking of dipping snuff and incenses.

She replies, "Yes sir, Mr. Sherif, how can I help you?"

Mr. Sherif replies, "You're not ready yet, Miss Bishops?"

Elise replies, "My name is not plural. It's Bishop, not Bishops, but whatever, ready for what?"

You could see it in his eyes he was confused by her remarks. Sherif looks at Elise, slightly puzzled, saying to her, "Didn't you read your memo? I put it on the door yesterday morning."

Elise says to Sherif, "Memo. What memo?"

He says to Elise, trying to hold back his frustration. At that very moment he looks down and sees that his yellow memo paper was on the floor in the hallway, sitting next to the trash can. Then Sherif speaks in a forgiving tone to Elise, "I'm sorry about that. I guess it fell off your door."

Elise replies, "What's it about?"

Sherif says, "We are replacing all the hot water heaters in the building to save energy. The boss is going green."

Elise standing there, for a moment she paused, then says, "OH, ok. Well, as you can see, I'm not ready." So Sherif says to Elise, "I'll come back to you last on this floor."

Elise replies, "Thank you, sir. I really appreciate that. Sorry about the mix-up. I'll be out of here within the hour."

Sherif replies, "No problem, no problem, Ms. Bishop. You have a good morning."

Elise replies, "You do the same, Mr. Sherif." And she closes the door.

Elise walks away from the door, goes to the kitchen, and picks up the plate with her toast on it. She walks to the dining room table and takes a seat.

Curious to see what was inside the package she had just received. As she starts to open the package, the phone rings; she stops to answer the phone. In a frustrated manner she says, "HELLO, WHAT!" She says to whomever it was on the other line.

The person on the other line replies, "It's me, girl! Foxy."

Victoria Denise Robertson, code name Foxy Roxy, a.k.a. the Carmel Queen mean about that green. She is six feet, one inch tall. Her skin is smooth and dark, like a crispy caramel chocolate Hershey's bar, if you can imagine that. She is a marble of wonder; she is a living, breathing, walking, and talking conundrum. A unicorn, a true-blue caramel black bone. She has a very, very, very large natural afro and always wears huge jewelry. Nothing but the finest for her. She has a gold tooth on the top front left-side of her mouth. She looks like she just walked out of a '70s Vogue Magazine.

She really does not have any family that she knows of. She grew up in the system as an abandoned child, left on the steps of the Montgomery fire station just outside Houston, Texas, in Conroe, Texas. Then taken to an orphanage when she was two months old. She does not know who her real family is or where she comes from. She floated from foster family to foster family until she was seventeen years old. She does not think about her real family too much anymore. But she likes to believe it made her stronger.

She graduated from the Aldine High School in Houston, Texas. She also attended and graduated from Prairie View University with a bachelor's degree in science and English. She served four years in the air force, two years in the army where she learned how to use explosives, weapons, and has an intricate knowledge of Black Ops counterterrorism strategies, using Sun Tzu, *The Art of War*. She got a job working for the CIA when she was twenty-six years old.

She met Elise and Silva at the CIA Academy. They kicked it off almost immediately. They helped each other to stay focused until their graduation day.

Foxy Roxy became close to Elise's family. Soon after that, they considered her family.

Getting back to the phone call when Elise figured out who was on the phone. Elise's facial expression turned to instant happiness in a matter of seconds.

Foxy says to Elise, "Well, good morning to you too, bitch!"

Elise quickly responds to Foxy, "Aw, girl. I am sorry. Rough morning. I got a package in the mail. Every time I try to open it, I get interrupted."

Foxy replies, "Aw, my bad. Just calling you to let you know that me and Silva are downstairs waiting on you. The Boss needs us in early this morning; they say it's urgent, *girl!*"

Elise says, "It's that time now?"

Foxy replies, "Ah, yeah."

Then Silva speaks out, screaming into the phone trying to speak to Elise before Foxy hangs up.

Hearing Silva on the phone, Elise tells Foxy to tell Silva, "Good morning to you too, bitch."

Foxy face smirks with laughter when she turns to deliver the message to Silva. Silva heard her on the phone before Foxy could respond Silva replies, "Ditto, bitch."

Linda Kay Lopez, code name Silva Tongue. She is of Puerto Rican descent. She is five feet, ten inches, with a dark red tan, long straight black hair, and dark brown eyes. She is very fast-talking, MAMACITA, with a heart of gold.

Like Foxy, she too was part of the system. Only because her family was all killed in a boating accident. Her mother, father, aunts, uncles, and her brother Javan who was nine years old at the time. No one knows what really happened on that tragic day.

Silva was with her grandmother when the authorities walked upon her grandmother's porch to give them the bad news that her entire family was in a horrific boating explosion. The police investigation concluded that it was faulty wiring that caused the accident. She was five years old when this happened back in 1975. Her mother and father were in their early thirties when they had Silva.

Her grandfather passed away before she was born. He was seventy-five years old. The only thing she knew about her grandfather was what her grandmother had told her. He was in the army and that he fought in WWII.

Her grandmother took care of Silva until she passed away three years later. Silva was only eight years old. Silva has always believed her grandmother died from depression and that she could not bear the pain of losing her entire family. She believed her grandmother held on as long as she could.

All Silva remembers is she died on her deathbed with Silva holding her hand as she was surrounded by nurses and the chaplain at the hospital.

Silva remembers looking down into her grandmother's face, holding on tight to her. Her grandmother looked up into Silva's eyes, barely breathing, she tells her grandbaby, in a whisper, to give her some water because her throat was so dry. Silva, trying to hold back her tears as she turns around, saw one of the nurses standing behind her reaches to hand her a cup of water.

The nurse says to Silva, "Go ahead, baby. I know, I know. It is hard. It's going to be alright."

Silva looks up at the lady nurse with a mean stare, thinking to herself in a very rude and irate thought she better not say out loud. She grabbed the cup of water very aggressively, without saying a mean word to her. Silva's physical gestures toward the nurse was worth a thousand words.

The chaplain puts his hand on Silva's shoulders and says, "Ok now that's enough. Let's focus on your grandmother; this is what's important." Silva looks up at the chaplain, shaking her head in agreement and replies, "Ok, sir."

She helps her grandmother lean forward. She places the cup of water very gently on her lips.

Her grandmother takes a couple of sips and says to Silva, "Thank you, baby." Then she lays back in the bed and tells Silva, "You are to be the best at whatever you choose to be. Don't let anyone tell you that you are a failure or you cannot do this or that. Prove them wrong. Make me proud, baby."

Then she starts coughing very badly. Silva screams out loud, "No Granny, don't go!"

Her grandmother takes her last breath. Then the machine flatlines.

They try to console Silva, but it is no use. They hurry her out of the room to the front of the building where there is a middle-aged, short, skinny, blonde-haired white lady dressed in a green-and-white blouse, yellow pants suit, and wearing green-and-yellow stiletto high heel shoes standing by a car. She whisked Silva away to a Catholic orphanage where she stayed until she was seventeen years old.

Silva graduated from a public high school in Galveston, Texas. Then she joined the marines. She scored very high on the ASVAP tests. They placed her in the Officers Program. She graduated first in her class. She then became a candidate in the CIA Academy where she met Elise and Foxy. Instantly, they became the best of friends.

They graduated together. Little did they know they would become an unbreakable hard-hitting elite group of three the world has never seen! Going on countless missions together, they spent most of their free time together becoming part of Elise's family. Now let us get back to the phone conversation between Foxy Roxy, Silva, and Elise.

Foxy smiles in happiness, blurting out, "Well, *alright then*. Hurry yo ass up, girl."

Then Elise quickly says, "Coming bitches, be down in a sec."

Then Foxy and Elise hang up their phones seemed like at the same time.

Then Elise realizes she has no time to open the package or eat her toast. She puts the package down and quickly begins to get ready for work. Thinking to herself, I will check *the* package later.

Elise rushes to the elevator door, leaving the 10th floor, without a moment's thought. Elise gets on the elevator and quickly hits the 1st floor button. Before the elevator door could close, a neighbor gets into the elevator. A neighbor she dreads sharing air with. A lady by the name of Mrs. Quartermane.

Mrs. Quartermane is an eighty-year-old white woman with snow-white hair. She has very pale skin like she never stood a day in the sun. She always wears extravagant smock-like dresses, which gives off loud colors that seem to shine when looking directly at her. She walks her dog it seems like every hour on the hour. Her dog is an overweight white poodle, with painted pink paw nails and pink ribbons on both of her ears. Her name is Fi-Fi, a very spoiled and annoying K9.

As Elise gathers her courage and temperament, she gives off a pretentious, fake-making act; with every inch of her being, she turns toward Mrs. Quartermane and speaks.

"Hello, ma'am. I mean, good morning," Elise says.

Mrs. Quartermane did not reply to Elise.

Elise looks at her waiting for a response. Then speaks again, "Good morning." Giving her the benefit of the doubt, thinking to herself, maybe she did not hear me.

Mrs. Quartermane turns to Elise and replies, "Miss Bishop, your mail was in my mailbox again. What address are you giving to your people?"

Elise gasps, trying to hold her attitude in check. She knew Mrs. Quartermane has issues with everyone. But before Elise walks away, she replies, "My people, huh? Well, the mailman *is* my people. It's a conspiracy we are going to fill up all the white people's mailboxes and make you all pay for it."

As the elevator reaches the bottom floor to the entrance of the apartment building, Mrs. Quartermane replies with a grunt, "That's not funny," as she quickly steps out of the elevator.

Elise chuckles to herself. In satisfaction, she leaves the elevator heading to the door that leads to the front of the building where Foxy and Silva await.

Elise gets to the front door where the doorman "Fitts" is standing. He is known as Fitts, no first name, just Fitts. That is what everyone calls him. Fitts is a six-foot, one-inch round man of Creole descent. He is in his early fifties and is very soft-spoken. He is always helpful when it comes to sharing information on where to find what you need. Fitts is your man.

Fitts says to Elise as he opens the door, "You have a wonderful day, Ms. Bishop."

"You too, Fitts," Elise replies.

Elise gets into the car. Foxy has a blue 2020 BMW.

Foxy pulls off. Silva shouts out, "WHAT UP, BITCH!"

Elise replies, "Good morning, Ms. Silva." Then she paused for a minute, then shouts in a very loud manner jokingly of course. "What up, bitch!" very loudly.

Then Elise asks Foxy, "Why do you have that nitrogen canister attached to your side like that all the time?"

Foxy says, "Some bitches like mace to burn a muthafucka's eyes out. Although I really do not need it, I just like freezing their eyes out!"

Elise says, "You know that's some psychopathic shit, right?" Then Foxy replies, "Aren't we all, my sisters?"

Then everyone laughs.

Foxy says, "So bitches, donuts or muffins this morning?"

Almost simultaneously, they all yelled out, in a playful happy manner, "Mrs. Chan's-Delight Donuts on Lorraine Street." Mrs. Chan is just north of downtown Houston, off I-59 North freeway.

They reach the donut shop fifteen minutes later. The girls place their order and pull to the window. There standing in the drive-thru window was *the* lady herself, Mrs. Chan.

Mrs. Chan is a very short woman, about five feet, one inch, and always walks with her left leg sliding across the floor, as if she had a stroke some years ago. A very mild-tempered, slightly obese woman who has a gracious and uplifting attitude. She always speaks in a light and friendly tone.

Her hairline was so straight, it was like it was cut with a razor across the front of her face, and the back of her hair hung just past her shoulders. She wears very small glasses with a thin gold trim that seem to sit on the tip of her nose. No matter what she did or how she moved, they never slid or moved off her nose. Never once did they move, as far as Elise could remember.

She has been a part of the Fifth Ward Community for years. Her parents moved here after World War II. They entered the US on a broken-down ship in New York City, according to Mrs. Chan's account recollections in her story. Then they moved to San Francisco, California, to be reunited with existing family members from China. The Chan family bought a pastry shop in the early 1940s in Chinatown just up the street from the Watts Community that was, at that time, mostly colored.

Mia Chan was born in 1946. She got married at eighteen years old and moved to Texas with her husband, who was a scientist. He passed away five years after relocating to Houston, Texas, from a work-related accident. She never remarried.

Now getting back to the girls in the drive-thru.

Everyone speaks, "Good morning, Mrs. Chan."

"Hello, ladies, headed to work?" replies Mrs. Chan.

Everyone replies at once, "Yes, ma'am."

Mrs. Chan giggled. She gave them their orders. Then she says to Elise, "Baby girl, tell your mother I would like to see her when she gets better."

Elise looks in surprise and replies, "Huh, she comes to your shop every morning for coffee and orders two glaze donuts for the past fifteen years. She has never missed one day of coming to your shop."

Mrs. Chan replies, "I know."

Elise says, "When was the last time she was here?"

"Four days ago," replies Mrs. Chan.

Suddenly, Elise becomes puzzled. She thought to herself, my mom left a message on my refrigerator. I checked the date on the letterhead, and it was also four days ago.

Elise replies, "Ok ma'am, I'll let her know."

Mrs. Chan waves and the girl's wave back as they pull from the drive-thru window.

Foxy pulls off and looks over at Elise, "What are you thinking about, Elise?"

She looks up at Foxy and replies, "It's probably nothing. After we check in at work and finish our briefing, I need to give my mom a call."

Silva says, with great concern in her voice, to Elise, "When was the last time you spoke to mother, sis?"

Elise thinks back, "Well, I know five days at least. I had a 3-day training OP, came back to headquarters, and had to turn around on a day-and-a-half job. Siberias would not take anyone else for the job, at the time. That's when you and Foxy had that thing in Israel. So, I took that mission."

Foxy replies, "Oh yeah, I remember that little hostage thing you did over there in Costa Rica."

"Yeah, that's the one I took because you and Silva did that back-to-back thing also in Yemen. I just got back at 10:00 p.m. last night. I read her note on my fridge. This morning dated five days ago," Elise replies.

CHAPTER 3

The Morgue

Elise, Foxy, and Silva finally make it to their destination. As they pass thru security, they all go check in with the station chief for assignments and a debriefing.

Section Chief James Williams, five feet, eleven inches, 230 pounds, borderline obese but fairly nice-looking. A big boned man with dark skin. He has a mini afro, always wears Armani suits. He always smells like Issey Miyake cologne and coco butter. He has a go get them, get the damn job done attitude. His primary and only concern was making sure every mission is a solid success, to say the least.

Chief Williams is a die-hard American patriot and stays close to his family morals and values. With a conservative background, he is every mother and father's dream of what a son should grow into. A natural born leader, the full package from what everyone can see. A patriotic solider with a purpose. Loyal only to family and country.

They finally make it to JW's office, they all call him Boss. Boss waved them in and asked them all to have a seat. By him being so nice, his tone brought red flags. The ladies thought to themselves, *what in the hell is going on?* They all complied and took a seat. Waiting in suspense as to see what was coming next. He shuts the door and closes his blinds, to reinforce privacy, for the moment.

He walks over to Elise and grabs her hand. JW leans over to Elise and says, "Elise, I'm sorry. I do not know any other way to tell you this. Your mother was missing for over four days; she was found dead on the sidewalk in front of her home at four this morning."

Elise jumps up! Blurting out, "OH MY GOD! What are you saying?" She was engulfed with sadness and unbearable pain. It felt as if a mule had back kicked her in the chest!

Foxy and Silva reacted in disbelief and shock as they both reached for Elise grabbing her pulling her quickly into their arms in a loving and gentle manner. Both Foxy and Silva knew Elise's mother for over ten years and were distraught just as well!

They met Elise's mom at the secret graduation ceremony of the CIA. There were no videos or cameras allowed at the event. They had no pictures of themselves; because of the integrity of the department, they were sworn to secrecy. Given only a slight consolation of having a small group of vetted family members to attend the ceremony.

Foxy and Silva had no family of their own. They were raised in the foster care system as stated previously in the beginning. They adopted Elise's family as their own creating an unbreakable bond. A sisterhood dating back for more than ten years.

Behind the curtain, sort of speak, Foxy, Silva, Boss, and others inside the office also had a special connection with Elise's family. This is what they did. They tethered themselves to the most precious gem; they fought for love of country and family. This is what drove them to being the most feared country in all the world. Their goal was so vivid. Powered only by living thru each other's families, as this was the only thing that kept them grounded reminding them of their true purpose and what they all stood for! A beacon of justice, a symbol of what drove them to make a difference in the world.

They were fighting for the most precious privilege of all, which most Americans take for granted. Everyday freedom of speech, freedom to speak your mind, freedom to have power over your own body, and the freedom to decide your higher power. Giving you the right to control, the privilege to control your own life's outcome, a belief that is shared by all who would join or was chosen for the greatest honor of all. The purpose of doing right by others who need it. Who cannot fight for their own rights but still counted all together, we call it the USA!

Elise calms down; her face soaked with tears of pain, sorrow, and fury! Foxy and Silva still standing around Elise in a huddle. As she thought to herself, who could do such a thing to my sweet, sweet, gentle mother.

Then she lashes out again, in a bloodcurdling scream, talking to herself, as if she were talking to the person responsible, "*MOTHERFUCKER, I will FIND your ass! You son of a bitch!* Believe that shit! OH WHY! OH WHY! DAMN! DAMN! DAMN!" trying to catch her breath gasping over and over again.

Silva, Foxy, and Elise continue standing in a huddle desperately trying to comfort each other from the pain.

They all try to calm down. Boss responds saying, "I can't imagine what you're going through. But officially, we do not have permission from the heads of state to do anything. At this moment. But unofficially, we are at your disposal."

HE calls Elise's name out loud for everyone to hear, "ELISE. Baby girl, I feel like I lost a sister. Whatever you need is yours! I am putting you three on a three weeks' paid vacation. Pending the investigation of your mother's untimely death. Is this clear?"

They all face each other. They all say in unison, "Ok." They all quickly speak out, as if it were a contest and the quickest one to respond would get a prize.

But in their minds, their deep, dark, cold, vengeful thinking minds, they were all looking for revenge. To add fuel to the fire, they were all trained in the deadly arts of Ishi-o-him. The most ancient of all deadly arts, it is said to have been created by the heavenly Father himself. They also were experts in military hardware, which includes classified and unclassified weapons. With a special feel of responsibility to their way of life, liberty, and justice to a normal American eye, they would be undeniably true-blue American patriots hands down.

Elise gladly accepts the terms and says to Boss, "I'll call you when I need you, sir." She painfully ends the conversation with the Boss. Elise says, "Right now, I have to go see my mom and contact my family to let them know what has happened."

The Boss responded, "Ok, as it should be. She is at Southwest Memorial Hospital."

Elise replies, "Thank you, sir. I will call you when I need you."

Elise, Foxy, and Silva leave the Boss's office. They walk past their co-workers' offices. The entire office could see the pain on their faces.

One by one the office staff approached the girls and gave their condolences, "We will get those sons of bitches!"

A promise that was made by everyone in the office. They made a pact to themselves not to stop until the murderers were found! The elevator doors open. The ladies get in as they are followed by everyone's eyes in the office.

They finally reach the parking lot.

Foxy says to Elise, "I speak for me and Silva when I say this. Sis, your t-jones was more than a mom to us; she was refuge for us, you know. Silva and I, she gave us a reason to believe in a life again. So, believe me, sis, we going to ride with you to the end! Believe that shit girl! I love you, ok!"

Elise says, "I know, girl, without question, but right now let us get to the hospital. I have to ID my mother."

Silva and Foxy look at each other and then Silva says, "Ok sis."

Then Foxy says, "Whoever did this is walking dead and don't even know it!"

A quick remark flies out of Elise's mouth, "You GOT DAMN SKIPPY, those bitches have no idea what's coming!" Then Foxy peels off in her car.

They ride to Southwest Memorial Hospital off the I-59 Southwest Freeway in Houston, Texas. Just a few miles away from Sugarland, Texas, where Elise's residence is located. They make their way into the hospital parking lot thirty-five minutes later. Soon after their arrival, they make their way inside the main entrance of the hospital, they took, a look around. Silva spots a security guard making his rounds. She quickly approaches him and cuts off his path saying, "Can you help us?"

In surprise, the security guard looks toward Silva, then replies with a friendly gesture, "How can I help you?"

Silva replies, "Can you direct us to the morgue?"

The guard asks Silva, "What is your business there?"

Foxy pulls out her badge and says, "THAT'S classified, mister." As she looks down at his name tag which read, James Collins, Foxy continues, "Mr. James." Foxy was so frustrated at that she didn't point, she didn't even respect him enough to call him by his last name.

By that time, Elise and Silva both flash their badges at the same time.

The security guard, in awe, finally speaks. In shock he replies, "Ok, I am sorry, ladies. I was expecting government looking *types*."

Silva says to the security guard, "What do we look like, secretaries?"

The guard learns quickly that he has offended Silva. Mr. Collins speaks up in a calm, humble, and very apologetic way, "No! No! No ma'am. Like beautiful exotic models." Hoping that remark would relieve the tension in that moment, but it did not.

Foxy says, "Yea right!" now you sound like a damn male chauvinist.

Elise tells the guard, "Brother, stop while you're ahead!"

He quickly takes heed to Elise's advice as he spews, "Sorry. Sorry." All the while waving them to walk with him in a calm and professional way. He escorts the ladies downstairs to the morgue. Mr. Collins finally brings them down a long hallway to a door next to a flight of stairs. They go down two flights of stairs to a corridor that shares three hallways marked D, E, and F.

Mr. Collins turns around, gently speaking to the girls, "This gentleman will take you the rest of the way."

The girls gaze into the darkness. A tall black man came out of the shadows. He came around the corner of hall D. He spoke to the ladies saying, "Hello, Ms. Bishop." Then gestured to the ladies saying, "Ladies, my name is Detective J. B. Skaggs. I am the lead detective assigned to your mother's case."

Detective Skaggs is a fifty-four-year-old man. Tall, dark, and slender. He has a lanky walk, big round eyes, and straight grey hair. He has a very razor-clean haircut and goatee. He has a slight gap in his front teeth, a deep voice, and speaks with a lisp. He kind of reminds you of a bougie uncle that thought he was better than everyone in his family but always tries to play it off like he doesn't. He dresses in slacks and a long white sleeve shirt, which is rolled up most of the time, with a black tie and no jacket. He had smooth, black, old-timey Stacy Adams on his feet.

Mr. Collins greets Detective Skaggs, a.k.a. John Bryan Skaggs, with a handshake and says all in one breath, "Hello and goodbye, Detective Skaggs."

Then Mr. Collins politely looks at the girls, with a sigh of relief expression, and says, "Have a nice day." With an aggressive quick walk, without looking back, he leaves to finish his rounds.

Elise speaks to Detective Skaggs, "Where is my mother?" in a painful voice.

He escorts the girls to Elise's mother's body to be identified.

Elise walks closer and closer to the room. She becomes more and more agitated, plagued with an avalanche of confusing emotions, one mostly being pain that surges throughout her whole body. Then she becomes numb all over. Elise enters the room; Foxy and Silva follow closely behind her, dreading these last moments before the moment of truth.

Elise looks up and across the room. She sees a body lying on a stainless steel table. Completely covered in a white sheet. Detective Skaggs walks to the body. As he looks toward Elise, he gently motions Foxy and Silva to hold her and bring Elise closer. Reluctantly, they do so.

Elise started to cry louder and louder the closer she got to her mother's body. She finally reaches the table.

Detective Skaggs apologizes again for her loss. He then removes the sheet uncovering the face.

Elise looks down, then reaches for the hand of her mother. Elise has a flashback of the way her mother used to look. She was dark-skinned, beautiful, five feet, eight inches thick woman. She wore her hair in a mini afro. It seemed like she always wore a kitchen apron. She always dressed conservatively, very ladylike. She was extremely intelligent and always instilled high morals and values upon Elise and her little sister, Kawanna. She taught them to think of others before themselves.

In death, her mother looks exactly the same. Elise thinks to herself, she looks like she's sleeping. Almost immediately, she notices that her mother has strangulation marks around her neck. At least that is what it appears to be to Elise. She keeps it to herself.

She slowly caresses her mother's cold, stiff face. Then gently bends over and kisses her mother on the forehead and whispers in her ear, as if she could hear her even in death.

Elise says, "I will find out who did this to you, Mommy. I promise! I will send every last one of them to hell! Whoever is responsible, may the HEAVENLY FATHER have mercy on their souls, because their asses belong to me.

Elise stands up slowly; she turns and starts to walk away out of the morgue. Detective Skaggs tries to talk to her about filling out the necessary paperwork in order to release her mother's body to the funeral home.

Elise looks back at him with pain and rage in her face and in her eyes. Detective Skaggs sees the look and quickly volunteers to do the paperwork for her.

That is when Detective Skaggs sees that Elise was no ordinary woman.

Elise shakes her head up and down, then turns around and storms out of the room. Foxy and Silva look back at Detective Skaggs.

All they could see in his facial expression was surprise and shock. Almost instantly he hears Foxy and Silva spouting profanities under their breath at him as they exit the room but not before quoting a feminist haiku.

Silva says, "We'll take care of the funeral arrangements when we're done doing what we need to do, sir."

Det. Skaggs politely replies, "No problem, just let me know if there is anything I can do. Please don't hesitate to ask."

Foxy and Silva both speak out in a whisper saying, "Hell has no fury like a woman scorned!"

The ladies finally make it back to the parking garage. Foxy goes and retrieves her car from the parking attendant. She pulls up to Elise and Silva; they get in the 2020 BMW and leave.

Elise speaks softly and says to Foxy, "Let's go to the SPOT right now."

Silva says to Elise, "You know we never go there unless we have an OP." Then Foxy turns and looks back at Silva, then Silva says, "You guys know I be trippin' when I'm trippin'."

Elise and Foxy Roxy both smile at Silva in agreement but with love on their faces.

Foxy anxiously drives to the spot. They take I-10 west of Houston, headed to the outskirts of Bryan College Station, Texas, to an undisclosed location.

CHAPTER 4

The Booger

Now back at the police headquarters in Houston, Texas. Detective Skaggs is rushing to get to an urgent phone call. He sits at his desk and answers the phone. Screaming out loud to his colleagues, "I got it, guys!" Everyone looks at him. They were not moved at all. The usual day-to-day, mind blistering, sometimes great, sometimes not-so-great routine. Everyone carries on their day as if the moment had never happened.

Detective Skaggs sits down at his desk; he speaks into the phone saying, "Hello, this is Detective Skaggs."

The voice on the other end of the phone, sounding like an old church-like woman.

What disturbed Detective Skaggs was the very soft-like tone, which invaded his moment at the time and seemed psychotic.

The soft voice on the other end of the phone says, "Is that you, JB?"

Detective Skaggs replies, "Yes, Momma, it's me."

The voice replies, "Well, what do you have to tell me, right now?"

Detective Skaggs pauses for a moment, in a soft tone and shear fright he replies softly, "I have everything in order. The phone lines are encrypted with the chip you gave me, and I alone have access. No one knows a thing."

The voice replies, "Good, sweet baby boy. Like I know you would. You coming over to visit later, like I hope you will?"

Detective Skaggs says in a half-dead reply, "Oh yes, Momma. I will be there with bells on. I cannot wait. You know I love you, right?"

Before he could finish his sentence, the phone goes click. Followed by a recording of an old white lady, sounding like she is from the 1950s era saying, "If you would like to make a call, please hang up and try your call again."

Detective Skaggs crunches back in his seat, acting very relieved that phone call was finally over. He gradually musters himself to get up to his feet. Slowly he walks out the door of his office. He felt sweat beads falling from his head. It felt like hot, hard, stiff blood. Just like it felt the first time he was beat up in the first grade.

He could still feel the pain from that first punch he got in his face by a neighborhood bully, an older boy called *Booger*. He was about three years older than he was and still freakishly big for his age. He liked picking on younger, smaller boys then. It gave him great pleasure because he was teased by his peers in school.

The reason they called him Booger was because he was a big ole red boy. Like a Frenchman you know from Louisiana. A Creole boy looked like he could pick up a tractor and throw it a thousand yards. He was a very ashy person, very dirty in appearance, wrinkled up clothes and a dried up booger that hung from his face—it looked like it was a part of him. Like a body part. It seemed like it was always staring right at you, according to Detective Skaggs. At the time when he was a ten-year-old boy. The boy everyone called Booger, seemed like to young Skaggs. Booger kept a fresh pack of boogers all the time going up and down, always hanging out of his nose. He wondered to himself, how could he breathe through his mouth. That little green, balled-up booger seemed to never quit. It was almost hypnotizing. Just to look at it, swinging back and forth.

Hearing his mother shout a gesture at little JB saying, "You better not stop fighting. No matter what, boy!" Little boy Skaggs had been thrown down and half beaten. Although to most it seemed like he was afraid, but truly young Skaggs had not a scary bone in his body. He really was freaked out of his mind. He did not want to get that wet, icky, slimy green booger on him. It seemed to him he would rather fall to his death but not at the expense of his mother's wishes. Something in him clicked and little JB began to get up and started roaring like a lion. He jumped to his feet hoping he would not get icky green slime on his fist. Young JB started swinging. Call it luck, call it chance, but whatever it was it played in little boy Skaggs favor.

He hit Booger, with a Hail Mary punch, and connected. He knocked Booger off his feet; in surprise, everyone in the neighborhood screamed in excitement. Joy completely rushed in like a breached dam. Young JB Skaggs felt like a superhero. He thought he was on top of the world when he took Booger down. Booger came to his feet for round two. Lady luck was not on his side that day. Booger's mother snatched him up before he could witness his fate. Young JB Skaggs prepared his self for round two. When he opened his eyes, Booger was gone! Off he went. Booger's mother was so ashamed of what he had done. They had just moved into the neighborhood. He had embarrassed her; she was appalled by Booger's actions. She dragged him home to be punished. No one spoke of the incident since.

JB becomes a hero in his neighborhood for not backing down from a bully. That's what gave him his courage.

Thinking back to when he was a kid, he quickly remembered in that moment. That's what made him the man he is today. Making that act his legacy to vow to Almighty God that he would protect *the little guy*.

Now he has become a seasoned vet in the force. He thought to himself, what he has learned made him kind of doubt his abilities for finding justice after having that conversation with the lady named Momma.

Then he snaps back out his childhood delusions and comes back to reality.

Now Momma is on his ass.

He thinks to himself, how am I going to play this shit storm off. He puts both of his hands on each side of his head, as if he were about to pray. Making a grunting noise then, he looks around the office and continues to do his work.

Anyway, the girls arrive at the spot finally.

CHAPTER 5

The Call

Foxy gets closer to the spot. They leave the city far behind them, finally to be in GOD'S country. Beautiful green grass as far as the eye can see, the forest and cattle, horses, sheep, jackrabbits, wild raccoons, possums, and every other wild thing that nature has to offer was right there in rural Texas, USA. The girls ride through the countryside trying to relax a little.

Elise started to make some more calls. First, she dialed the guys that could possibly know her sister's whereabouts. Kawanna's closest friends, the Mad Pack. They all said they have not seen her in two days.

Elise was distraught. It felt like another blow to the gut and no closer to finding her sister.

Frantically, she continues to call Kawanna, in order to break the bad news about their mother, again and again. But Elise still gets no answer.

Elise calls the dean at the university. The dean's secretary answers the phone, "Yes. You have reached the dean's office. How may I help you please?"

Elise pauses for a moment.

The secretary saying again, "Dr. Johnathon Dornn's office. May I help you please?"

Elise snaps out of it. Quickly speaks into the phone, "I'm sorry, ma'am. My name is Elise Bishop. I have been trying to reach my little sister Kawanna Bishop all day. She is a student there. So far I have been unsuccessful in trying to reach her. There is a death in the family. Our mother has passed away this morning. Could you send someone to her dorm room to check on her for me? I would really appreciate it, ma'am."

Ms. Carter says, "Oh my GOD!" The phone goes silent for a couple of seconds. Not knowing how to respond.

The secretary finally answers Elise, "No problem, ma'am. We will send the message through an aid to her dorm, to inform her that family is trying to contact her. Is there anything else I can possibly do for you, ma'am?"

Elise replies, "No, thank you."

Ms. Carter says, "I'm so sorry for your loss."

Elise replies, "Thank you."

In that moment, a single tear rolled down Elise's left cheek, thinking about her mother again.

Ms. Carter says, "Ok, goodbye ma'am and God bless."

Elise hangs up the phone wondering to herself, *where the hell is my sister?*

CHAPTER 6

The Spot

Elise looks over at the girls after getting off the phone with the university. They could see it in her face that it wasn't good news. So they purposely made no comment; they remained quiet.

Foxy finally pulls into the driveway entrance to the spot. Elise gets out of the car. She looks around to see if there are any changes to the spot. She quickly puts the combination into the lock that is attached to the cattle guard. Foxy rushes to drive thru. Elise shuts the cattle guard bar behind them, then she locks it. They drive three miles down a dusty, white dirt road to the spot, where they come up on a red ranch farmhouse.

At first, it looked abandoned to the naked eye. Elise and Silva exit the vehicle. Foxy stays inside the car. Elise says to Foxy, "See you inside."

Foxy's reply was not verbal but a gesture. She nods her head up and down once to show that she complies. Foxy pulls the car into a big, red square-like barn. The barn stood about 35 feet high, 450 feet long, and 150 feet wide; you could tell there was more than just parking inside the barn. Everything was covered with military camouflage designs, hiding large and small items throughout the barn.

Foxy shuts the barn door and locks it down with a series of locks and latches. Topping it off with two dead bolts and a combination lock. She thinks to herself, yeah now that is done. Ain't shit getting in here, not even a bird could sneak a shit without her cheeks splitting toward her beak.

Foxy admiring her work on locking down the barn, she walks off thinking to herself, again appreciating the secure barn. It has no windows or back doors. She makes her way back up to the ranch house to meet with Elise and Silva which is about 60 to 70 yards away always. Enjoying her brief but very satisfying nature walk back.

Meanwhile, back at the dorm room.

CHAPTER 7

The Dorm Room

Monday evening, back at the University of Stanford, California, the dean was in his office. He is a noticeably short Irish man with fiery-red hair. He has a receding hairline. He stood about five feet, four-and-a-half inches. He looked very athletic and he always wore the finest suits he could find.

The faculty heard a rumor about the dean. No one really knows if it is true or not. It is said that he has a suit for everyday of the year, with accessories to match. He is fifty-eight years old. His name is Johnathon Dornn. A very generous, humbled-minded, by the book, one hundred and ten percent fair man. He has a very well-rounded personality.

He comes from a good family up north, in a place called Lived, Idaho. A very small town in the middle of rural Idaho, USA. His mother was a stay-at-home mother. Father is a farmer. No brothers.

His sister was adopted at the age of five. Her name is Julie Dornn. She is a secretary for the US Pentagon. Mr. Dornn's friends and family call him JD

Dean Dornn's secretary picks up her office phone to call Kawanna's dorm room in building 5 to let her know family had been attempting to call all day.

Someone answers the phone; there is a loud, moaning voice saying nothing into the phone. Ms. Carter says, "This is Ms. Carter, the secretary of Dean Dornn's office!" She replies in an offended remark.

Still hearing nothing but strange noises on the other end of the line, assuming it was a female. She says again, "I beg your pardon, little lady. Who is this? Tell me right now, this instant!"

The mystery sounding individual that answered the phone call was still moaning, then the phone hangs up.

Ms. Carter, a sixty-three-year-old lady has been working for the Stanford University for over thirty-three years. She worked for four deans and Dean Dornn makes number 5. She is a tall, skinny old Caucasian lady. She looked and dressed like a Mormon. Very, very, conservative. She had green eyes and snow-white Shirley Temple curls. She looked like she came out of a *Leave It to Beaver* episode. From her head down to her nun-look-alike shoes covering her toes. Her full name is Nancy Ann Carter. Never been married, no kids, and no brothers. Just an older sister she visits twice a year. Just outside Houston, Texas, in a retirement home in Woodlands, Texas.

Ms. Carter slams her phone down. Screaming to herself, in a whisper, rocking back and forth. Ms. Carter catches herself, calms herself down for a moment. Then she picks up the phone, clears her throat, and tries to reach the dorm building again.

No answer. The phone rings and rings and rings. Ms. Carter hangs up the phone, uttering profanities, as she whispers to herself, hoping to catch the individual that had just disrespected her on the phone moments ago. There was no answer.

Ms. Carter then quickly called down to the security shack, which is not too far from Kawanna's dorm building. The phone rings and rings; just as Ms. Carter was about to hang up the phone, a voice on the other end spoke to Ms. Carter.

"Hello! May I help you?"

Ms. Carter says, "Hello."

Then the voice replies, "Is this Ms. Carter?"

She replies, "This is Ms. Carter from the dean's office ..."

The voice replies, "I'm Ralph, ma'am. I know your voice, Ms. Carter."

In an instant her remarks shot from the phone like a bullet, it seemed to Ralph. Ms. Carter says, "I need you to get over to the girl's dorm building number 5. To see what the heck is going on …" Then she hears a heavy breathing moan, coming from the other side of the phone.

Then she hears Mr. Ralph replying to her, "I'm getting ready to go over there now, ma'am."

Then Ms. Carter added another request by saying, "And find out who is playing on the phone."

Mr. Ralph Gomez agreed so graciously to everything she was saying. So, with a verbal expression of gratitude, he continues, "It feels good to be needed. Thank you for my job." *I will kiss your ass, wherever and whenever, type attitude, sprung from his demeanor.*

Mr. Ralph says to Ms. Carter, "On my way, ma'am. Don't worry. I'll call you right back with an update, ma'am."

Ms. Carter says, "You call me RIGHT back, Ralph." Saying it twice just in case he didn't understand, she thought to herself.

"Yes ma'am, you can count on me," Ralph replied.

Ms. Carter, appalled by his Mexican dialect, misses most of what he was saying. She says, "Ok, whatever," then hangs up the phone.

Mr. Ralph is obese and of Hispanic descent. He is always eating something. Very thankful about everything. He stands about six feet, five inches and weighs 345 pounds. A pussycat wouldn't hurt a flea type of personality. His size is just a fluke. He is an illegal alien that turned out to become a legal citizen of the USA, who happened to be in the right place at the right time.

When a thirteen-year-old blonde hair, blue-eyed, Caucasian boy was crossing in a crosswalk, downtown in Palo Alto, California, Ralph was getting off a city bus downtown coming from a failed attempt to find a job. He and a crowd of people got off the transit bus in droves trying quickly to catch the next bus for transport to their final destination. The little thirteen-year-old boy was listening to his IPods. He didn't look both ways obviously; if he had, he would have noticed the truck coming right at him. Then he continued to proceed to walk across the street. When suddenly, the truck came closer barreling around the corner.

In no time flat, Mr. Ralph got off a bus, opposite side of the boy. He had passed the boy in the crosswalk. He spoke to the boy; the boy didn't notice him.

Mr. Ralph saw a blue F-150 truck recklessly speeding through traffic. Soon after the speeding truck, there was a motorcycle cop blazing right on his tail.

Mr. Ralph looked back. The boy did not hear everyone screaming at him to get out of the way. Without any regard to his own life, Ralph stretches out his arms to grab the boy. Still the boy oblivious to the ongoing dangerous event that was occurring. By some miracle, the boy finally looked up, and in an instant, everything to the little boy went in slow motion.

Then the boy pulled down his earphones saying, "What?" He looked over his shoulder. All he saw was the front end of the blue ford F-150 truck coming straight for him. The boy gave in and with great courage the boy embraced his fate. Then suddenly just before impact, a body came out of nowhere and tackled him out of the way. Before the boy knew what happened, the truck and the motorcycle cop were long gone. Out of sight. The driver was apprehended twenty minutes later on the freeway by the highway patrol.

The motorcycle cop saw the heroic action of the man who risked his life to save the little boy. He had the whole event recorded on his police patrol camera, mounted on his motorcycle. The policeman took the tape to the news so the city would know of the stranger's heroic deeds. Soon after, Mr. Ralph was recognized and brought before the mayor to be awarded recognition for his bravery.

When they found out Mr. Ralph Gomez was an illegal alien, who saved the boy, the boy's father wanted to show his gratitude. The boy's father was a very rich and powerful individual in the city. He pulls some strings to get Ralph Gomez a work visa and a job at Stanford University, where Mr. Gomez has been employed for the last three years. Mr. Gomez has been so happy and thankful ever since.

Mr. Ralph hurries over to Kawanna's dorm room. Minutes later, he walks up inside the main entrance of the dorm building. He looks around, then he walks up to the information counter. He sees no one at the desk. Mr. Gomez became puzzled, thinking to himself, there is always a person at the dorm counter.

He knew that from the University Rules and Regulation Handbook that he read upon getting the job. In the back of his mind, he knew they were most likely doing mischief like he's caught them in the past, always letting them off with a warning. Mr. Ralph regretfully feels sad for the girls because he knows this time he has to write them up. Standing stuck in a trance of a daydream, he finally snaps out of it.

Mr. Gomez shouted out hoping to hear a reply, but there was none. He went toward one of the dorm hallways, trying to spot someone to figure out what was going on. He had questions he wanted answered. So, he walked and walked until he came upon the study lounge door to his right, which was locked.

Then he relaxed for a moment thinking to himself, the girls must all be watching a really good movie; that's probably why they left the counter unattended. As Mr. Gomez reached for the door, suddenly the lights to the dorm building go out.

Mr. Gomez was startled at first, then he hears mumbling and moaning and groaning. To himself he thought, the girls were cussing in their own little girl way. Mr. Gomez pulls out his flashlight. In a flash, quickly he flicks on his flashlight and what he sees shocked him into fright.

Trying very hard to believe what he was looking at was not happening. In great fright and a complete loss of the human language, in amazement feeling responsible for the girls, he remained focused in enough time to see the girls lying on the floor gagged and bound, trying to get loose. As he walked toward the girls, the girls got very loud and their eyes turned white with fright.

About the same time, Mr. Ralph understood what they were trying to do was to warn him. He quickly turned around; everything goes black for Mr. Gomez.

CHAPTER 8

No Leads

Three days later, Thursday night, the girls already heard about the invasion at the dorm on Monday night, and based on their sources, they knew Kawanna was not there. The ladies keep working trying to find Kawanna.

Silva and Foxy wanted to know why Elise won't activate Kawanna's cell phone, then Elise explains why saying that Kawanna and she had a pact that we would not invade each other's privacy because of who I am, and I agreed to it.

Foxy replies, "I know what you're saying, girl, but this is an emergency, girl. She'll understand I'm sure."

Elise says, "I know exactly what you're saying, but I have to explore every other option before I resort to that being last; please respect my decision ok."

Foxy and Silva both agree to Elise's wishes. Then they continue back working checking the airports, bus stations, hospitals, and anywhere else that has security cameras.

They even had their informants working day and night. Hoping to hear anything from the streets that might lead them to Elise's baby sister.

Elise says, "I can't believe this fuck shit. It's been almost a week and still nothing. Ah, man. Please, God, help me find my sister!"

Silva says to Elise, "I've been cross-checking Kawanna's phone records, and I could not find anything that stuck out."

Foxy says, "I also checked every hotel, motel, and bed and breakfast in Houston, Texas, and Stanford, California. Within a 50-mile radius of each city and still nothing."

Then a frustrated Elise tries to calm herself down by taking deep breaths. She turns her chair around facing Foxy and Silva and says, "You know, we need to stop now and get some rest."

Silva replies, "Girl, are you sure, we've only been up for two days." To Silva, considering the circumstances, that was nothing.

Foxy says, "Now Silva, you know we have to rest in order for us to be more effective."

Elise agrees with Foxy and shakes her head up and down.

They call it a night.

CHAPTER 9

The Witness

Meanwhile twelve to fifteen hours later, Thursday evening, Ralph was awoken by heavy breathing. Without warning, his eyes pop open looking like the cartoon character Shaggy on Scooby-Doo mystery show. His sight is blurry at first. He regains his sight in complete surprise. He notices a large crowd standing around him, then he hears a deep voice piercing through the crowd of reporters saying, "Everybody look! He's awake! Get the nurse!"

Mr. Gomez in complete surprise looked upon the people standing in his hospital room.

As everyone in the room noticed that Mr. Gomez was conscious, suddenly there was a flurry of shouting and yelling coming from the reporters trying to get the scoop on what exactly happened at the scene from the horse's mouth himself.

Then Mr. Gomez heard an even louder and very deep voice saying, "Everyone, if you could please exit the room so the patient can get some rest!"

The shouting stopped. Standing in the doorway was Detective Skaggs and Mr. Gomez's doctor.

Everyone complied but not without very disappointing moans and groans coming from the crowd. In perfect unison, they all say, "All right, all right."

Then they all cleared out of Mr. Gomez's hospital room.

His doctor was a tall, stick-figured man. He stood about six feet, five inches tall, wearing very thick bifocals encased in a black frame. He was of Caucasian descent. He had a bald-shaped peanut head. His name is Dr. Gibbs.

As Dr. Gibbs approached, he pulled out his eye light and pointed the light into Mr. Gomez's eyes, while asking him a series of questions in Spanish regarding his health. Mr. Gomez answers the questions assuring that he was of sound mind and feeling better.

As Dr. Gibbs finishes his physical examination, he turns to Detective Skaggs and gives him permission to question Mr. Gomez, then quietly exits the room, shutting the door behind him.

Detective Skaggs introduces himself as he walks closer to Mr. Gomez's hospital bed. Detective Skaggs leans over to Gomez and in a grinding deep voice says, "You gave us a little bit of a scare, big guy."

Mr. Gomez replies in a painful, aching, exhausted, scratchy voice, "Sorry, sir," with a very strong Hispanic dialect.

Detective Skaggs says, "Mr. Gomez, don't apologize. Do you know how long you were out?"

Mr. Gomez looks up, with a lost and confused expression and says, "Sir, I'm sorry. I'm sorry. I don't speak English very well. I'm still learning, sir."

Detective Skaggs leans back, whispering to himself. Stomping his feet on the floor, in a very disappointed manner but not forcefully or aggressively, he knew from his file he could speak real good English, but he didn't want to spook him, so he played his game.

He gathers himself with his back turned to Mr. Gomez. Detective Skaggs turns around saying, "Ok sir. I'll be right back." He quickly turns away to leave out of the room.

Mr. Gomez reaches for the remote control, then turns to the television, and then presses the button for the nurse in order to request something to eat and drink. Then turns on the television and finds a show on Telemundo and starts to watch the program.

Meanwhile, Detective Skaggs thinking to himself, while approaching the nurse's station, "I've learned French, Russian, and German but never took the damn time to learn Spanish."

When he approaches the nurse's station, he asks if anyone spoke Spanish. He hears a low, soft-speaking voice. It was a petite, very short middle-aged Mexican woman in her early fifties. She was with the housekeeping department, working at the nurse's station. Detective Skaggs looks around, only to find the woman looking back at him. In unison, all the ladies at the nurse's station quickly looked toward the same woman he was looking at.

Detective Skaggs looks at her name tag. It read Ramirez. He formally introduces himself saying, "My name is Detective Skaggs, Mrs. Ramirez. I need someone to translate for me."

Mrs. Ramirez replied, "If it's ok with my boss."

The doctor looked toward them and said, "It's ok, go ahead."

Mrs. Ramirez replies, "Ok sir."

Detective Skaggs replies in a friendly manner; with happiness and relief, "I did not realize you spoke English so very well. It is better than mine," he jokingly says!

Mrs. Ramirez blushes. She is four feet, nine inches tall. She is short and round. Has salt-and-pepper hair, brown eyes, and caramel brown skin. Mrs. Ramirez replies to Detective Skaggs, "Thank you, sir. How can I help you today?"

Detective Skaggs explains to Mrs. Ramirez what he needed her to do. They walk into Mr. Gomez's hospital room. Mr. Gomez sits up in the bed.

Detective Skaggs introduces his self again, "I am Detective Skaggs from the Houston, Texas Police Department." Mrs. Ramirez proceeds to translate for Mr. Gomez.

"Soy el Detective Skaggs del departamento de policía de Houston, Texas."

Mrs. Ramirez looks toward Detective Skaggs as he proceeds, "You gave us a little bit of a scare. You had an allergic reaction. You were out since

Monday night. You had a reaction from the ether that was used on you by your assailants."

"Nos diste un poco de miedo. Tuviste una reacción alérgica. Estuviste fuera desde el lunes. Por la noche cuando tuviste una reacción química del Éter que te usaron tus atacantes," Mrs. Ramirez continues.

Mr. Gomez replies, "OK señor. ¿Qué estabas tratando de descubrir?"

Mrs. Ramirez translates, "Ok sir. What were you trying to find out?"

Detective Skaggs says, "I am working with the Stanford Police Department. The case I caught in Houston, Texas. We think they are linked. Can you tell me anything that I can possibly use to catch these criminals, sir? The girls at the dorm for some reason have no memory of the event."

Mrs. Ramirez translates, "Estoy trabajando con el Departamento de Policía de Stanford. El caso que atrapé en Houston, Texas. Creemos que están vinculados. ¿Me puede decir algo que pueda usar para atrapar a estos criminales, señor?"

Mr. Gomez looks back toward Detective Skaggs and replies, "Vi hombres con pasamontañas negros, vestidos de negro. Entonces, antes de darme cuenta, me arrojaron al suelo y me sometieron. Escuché gritar a uno de los hombres, pero con un suave tono bajo. Parecía que estaba hablando por teléfono, porque le dijo a la persona por teléfono, no está aquí y la niña no está en ninguna parte, hemos buscado en todas partes, pero seguiremos buscando. El jefe, al menos sonó como si fuera el jefe para mí, continúa hablando. 'Mantenerlo vivo.' El hombre n.° 2 dijo. El hombre n.° 1 me empujó hacia abajo, cara primero, más cerca del piso. A medida que su agarre se hizo más fuerte, creo que era el hombre # 2. Se arrodilló en el suelo, muy cerca de mí, susurrándome al oído: 'Dígales que nos den lo que queremos y nos detendremos.'

"Mientras estaba tirado en el piso, su guante se volteó hacia atrás y vi un Escorpión Rojo en su mano entre su pulgar y su dedo índice. Grité a mi agresor en español, luego él me habló en español. Luego me agarró la cara y todo se volvió negro," continued Mr. Gomez.

Mrs. Ramirez translates, "I saw men in black ski masks dressed in black. Then, before I knew it, they rushed me to the floor and subdued me. I heard one of the men's voice yell out but with a soft under tone. It seemed like he was talking on a phone, because he said to the person on the phone, it's not here and the girl is nowhere to be found. We've searched everywhere, but we'll keep looking. The boss, well, at least he sounded like he was the boss to me, continues to speak. 'Keep him alive,' man #2 said. Man #1 pushed me down, face first, closer to the floor. As his grip got tighter, I think he was man #2. He kneeled down to the floor real close to me, whispering in my ear saying, "Tell them to give us what we want, and we will stop."

"As I was lying on the floor, his glove flipped back, and I saw a Red Scorpion on his hand between his thumb and his index finger.

"I screamed out to my assailant in Spanish, then he spoke to me in Spanish. Then he grabbed my face, and everything went black." Mrs. Ramirez translates everything to Detective Skaggs.

Then the head nurse walks in during the conversation warning Detective Skaggs of Mr. Gomez's stress level had elevated his blood pressure. And just like that the interview was over.

Detective Skaggs gestured to the nurse in compliance. Then thanked Mr. Gomez for his cooperation.

A tired and worn-out Gomez sinks down into his hospital bed in relief. He slowly closed his eyes saying, with his strong Hispanic dialect, "Ok, thank you. Goodbye, sir."

Mrs. Ramirez finished the translation and Detective Skaggs thanked her, then thanks Mr. Gomez. Then Mrs. Ramirez exits the room.

Detective Skaggs hurries to finish jotting down his notes on his black notepad. Then quietly leaves the hospital room. He is met by a tall, Spanish, dark-haired man in a police uniform with lieutenant bars.

His name is Jesus Garcia. Lt. Garcia is a very large slightly obese man. A very soft-spoken individual. As he approached Detective Skaggs, they greeted each other with a polite, pleasant tone.

"My officers notified me that you would be here at the hospital. So, is the victim conscious yet?" said Lt. Garcia.

Det. Skaggs replies, "He is in and out of consciousness due to the medication. I spoke to the victim very briefly telling me he could not identify the perpetrators and that one of the kidnappers talked to him, and I quote, 'Give us what we want and we will stop!'"

Lt. Garcia's reaction was puzzling. Then he said to Detective Skaggs, "I wonder what they meant by that?"

Detective Skaggs replies, "Your guess is as good as mine. I'll wait for your official report concerning the statements of the girls at the dorm. I have to catch my flight back to Houston, Lt. Garcia."

"Yes sir. I'll have those reports sent to you as soon as they are available. I have a car waiting for you in the front entrance of the hospital.

"Before you leave, I would like to ask you, off the record of course, are you, somewhat, associated to some of the victims? Because in my personal opinion, don't take this the wrong way, most offices I've dealt with in the past usually don't go to such lengths as to physically take themselves this far away for a home invasion with no casualties, unless he or she has some type of personal connection to the case. To me, that could only mean one thing," says Lt. Garcia.

Detective Skaggs looked at Lt. Garcia with a blank and shocked look on his face. As he thought to himself, "Damn it! He is onto me!"

Before Detective Skaggs could answer the question, Lt. Garcia screamed out in a loud, happy, satisfied tone saying, "Now that is what I call being passionate about your work. I used to be just like you when I was in homicide, back in the day. Well, I can go on and on blabbering about my past conquests. Let me walk you out to the front if you don't mind."

"Sir." Detective Skaggs graciously replies.

CHAPTER 10

The Farmhouse

Friday morning 10:00 a.m., four days after the assault on dorm 5, back at the dean's office, Mr. Dornn rings Ms. Carter.

She answers the phone. Mr. Dornn says, "Have you alerted all of the victim's families to assure that there will be a full investigation and they will be briefed as soon as we know everything?"

"Yes sir, I did, and sir, I just got off the phone with Chief Bale of the Stanford Police Department. He is sending over a representative to brief you about Monday night. As soon as he or she arrives, I will promptly alert you, sir," Ms. Carter replies.

"Thank you. Oh yeah, did you send flowers to Mr. Gomez and a condolence letter to the Gomez family?" asks Dean Dornn.

Ms. Carter says, "Already done, sir."

Dean Dornn replies, "Thank you again, that'll be all for now." He hangs up the phone.

Three days earlier, that Tuesday morning, back at the farmhouse the girls are trying to focus on the task at hand. Together the girls start to formulate their mission. Hoping to find any information that could explain the very painful, confusing, very insane shitty-ass events that were going on.

Elise, Silva, and Foxy are sitting at their stations, constantly typing and typing rapidly on their keyboards. Three days have passed and still no leads. After an understandably sleepless and body tossing night after night, barely eating, they continue to go at it again and again.

They are in a zone, oblivious to everything around them, fueled only by the passion of their love for Kawanna. Their fingertips are fat and swollen from the nonstop typing. Trying to find out anything that could bring them one step closer to finding their sister Kawanna. Then Foxy stops abruptly for a moment to gasp for air. Then she runs her fingers through her hair shouting out loud for all to hear. In anger and frustration, for a moment, she loses it and screams out "AHHHHHHHHHH!!!!!!!"

She catches herself; as she looks around, her eyes meet Elise's eyes. Quickly Foxy says to Elise, "Sorry, baby, you know. I'm just a little bit frustrated that I'm not nowhere close to finding Kawanna's whereabouts, and I'm angry and sad at the same time, sis."

Before she could speak another word, Elise cuts her off saying to both Silva and Foxy, "Sisters, I know we have been racking our brains out to the max. I appreciate everything; really, I do. Let us try to stay focused on the task at hand. Take your time, ladies, because, believe me, on my life and my soul to hell, we will find these muthafuckers that are responsible, believe that!"

Silva turns around in her seat agreeing. With a loud screaming voice, "You damn skippy, baby girl! That I do know." She jumps out of her chair, runs over to Elise, and threw up a high five! Their hands slapping together in a quick and fast manner.

Then Silva and Elise both turn toward Foxy. Foxy puts up both hands, slapping Elise and Silva's hands at the same time, up high quick and loud! Foxy says, "Shit, I'm hungry." Then Elise replies, "Your ass is always hungry, baby Huey." Then Silva says, "I'm hungry too. Who's cooking tonight?" Then Foxy replies, "I got it. It's my time anyway."

Elise and Silva both look at Foxy and put their order in. Foxy prepared their food for that day. Then everyone without haste return back to their work.

Elise reluctantly accesses her sister's cell phone and activates Kawanna's GPS tracker. Then the strangest thing happens. Elise stares at the computer screen. In shock for a moment, she then whispers to the girls. In disbelief, Elise says, "Fuck me!"

Foxy overhears her and jokingly she spats back at Elise, "No girl, I love you but not like that, bitch."

Oblivious to Elise's moment and what she had discovered, Foxy finishes by saying, "Girl, please we ain't got time to play."

Elise then recovers from her shock, stands up, and shouts out to the ladies, "No, for real, look at this shit!"

Silva and Foxy turn around in their chairs looking at Elise's computer screen. Then they react the same in complete shock.

Silva and Foxy say in a whisper, "Fuck me!"

Then Silva says to Elise, "Is this shit for real?"

Elise replies, "As far as I know, this is what the system says; I hope it's not a glitch."

"Bitches, let's go now. I mean right now!" says Foxy.

They quickly grab their plans and exit the farmhouse. Foxy goes to retrieve the car while Silva locks down the computers. Elise heads to the back room after transmitting the GPS information to her satellite mobile tracking device.

The girls exit the farmhouse, racing back to Elise's apartment building because the GPS device shows that Kawanna is in her apartment!

Elise starts dialing Kawanna's phone. The phone just rings and rings, then keeps going to voice mail. Then suddenly she gets a call on her second line. Not wanting to hang up the phone, just in case her sister answers, she sees that it is the dean's office at Stanford on the other line.

As Kawanna's voice mail picks up, to Elise it seemed like for the hundredth time, then Elise leaves another message saying to her sister, "If you are at my house, do not leave! I'll be right there! I love you very much."

Then without breathing the phone clicks, she answers the other line, speaking softly she says, "Hello."

The voice on the other line says, "This is Ms. Carter. Is this Ms. Elise, Kawanna's family?"

"Yes, I am, Ms. Carter," Elise replies in a moderate tone.

Ms. Carter asks Elise, "Can you please hold for one moment? Dean Dornn would like to speak to you."

"Yes ma'am, I can," Elise replies. Ms. Carter places the call on hold.

A moment later Dean Dornn's voice is on the phone. With a hesitant voice, he takes a deep breath then he says, "Hello, Ms. Bishop. I am sorry to inform you that prior to the incident at the dorm building your sister Kawanna is assigned to, looking at our records we've noticed that she's been missing for two days before the raid on dorm 5." Elise replies, "Raid dorm 5." Dean Dornn responds saying, "Yes, that is what the media is calling it. That happened Monday night, and it's been like a circus around here, but I want you to know we're doing our very best to get to the bottom of this, Ms. Bishop. She'll turn up, the kids here do this type of stuff all the time. I'm very confident that one event has nothing to do with the other. I'm sure of it, but in the meantime, if anything turns up, we'll contact you first thing, Ms. Bishop. I didn't want to alarm you until we have more to go on.

"A representative from the Stanford University Police Department has informed us that they are following upon some really good leads. My security guard on campus might be a valuable witness to the case. They say he's awake now; when they know more, we will know more."

Elise replies to the dean in an exhausted and worried tone, "Well, I activated her GPS tracker on her phone. I am currently tracking the signal back to my home."

In relief, Dean Dornn speaks out loud saying "Well, there you have it. Maybe she wanted to surprise you."

"Well, I sure hope so, Mr. Dornn. I'll let you know what happens, sir," Elise replies in a hopeful voice.

"OK, Ms. Bishop and I'll do the same." Dean Dornn hurriedly replies, with a relief in his voice. He continues saying, "My prayers are out to your family." Then hangs up the phone.

Foxy looks over at Elise, then she says, "So what's the word? Did we catch a break?"

Elise looks back at the ladies, with a disappointed expression she replies, "It's nothing. They haven't found or know a damn thang. It is up to us, ladies, to get to the bottom of all this shit!"

All agree and reply in a sadistic, revengeful way, in unison, both Foxy and Silva say, "Sho you right!" (Ebonics translation, we sure will, girlfriend.)

They get closer to Elise's apartment.

CHAPTER 11

The Talk

Late Thursday night, Detective Skaggs finally arrives back at Hobby Airport in Houston, Texas, tired and exhausted. He grabs his luggage, then he makes eye contact with a mysterious dark man; the dark man stood about seven feet tall, 300 pounds, very dark-skinned complexion with a face full of deep dark black hair, a black gangster hat resembling Al CAPONE's attire with a black leather trench coat and black-and-white snakeskin Stacy Adams shoes with black sunglasses on. He looked very terrifying.

The dark stranger grabs hold of Det. Skaggs's arm quietly and politely he guides him through the baggage claim area together.

As they walk through the airport exit doors, there sitting in the pickup area was a long black Lincoln limousine. The chauffeur opens the door; Detective Skaggs gets inside of the limo seated next to the dark stranger. There sitting inside was a shadow of a woman. She was sitting on the far corner on the opposite seat from Detective Skaggs. There was a small light coming from the dome of the roof.

The limo pulls away from the airport. Detective Skaggs speaks in a very soft-spoken voice to the Shadow Lady and says, "Ma'am, I got bad news and good news."

The Shadow Lady slams down her hand on top of the seat where she is sitting angrily! She says to Det. Skaggs, "Well, let me inform you that lead at the dorm was a bust, just nothing but bullshit all around."

Detective Skaggs quickly jerks in a defensive reaction, with his hands up toward his face.

The Shadow Lady says to him in a screeching, high-pitched voice, sounding like two coffee saucers rubbing against each other. With a devilish, aggravated tone, she says, "I'm tired of your damn failure. You're no good!" In mid-sentence she stops and catches herself. "Oh baby, I'm sorry. Did I scare you? Well, I'm sorry but you know how Momma gets when I am upset, don't you, baby?"

Detective Skaggs replies, "Yes, Momma. I know. I promise I will have better news next time. Please believe me."

As Detective Skaggs pleads, it seems like for his life, at the same time the car stops. She looks up at Det. Skaggs; he could see that begging and pleading were on deaf ears.

The Shadow Lady says, "I trust that you know my feelings about failure are not an option. I know what you've been doing. So, your services will no longer be needed."

At the same time, the door opens and her henchman grabs Detective Skaggs by his neck. To the henchman, it seems normal. Detective Skaggs's reaction was of horror. Detective Skaggs struggled to break free. The henchman pulled, then tightened up his grip around Detective Skaggs's neck, almost to the brink of unconsciousness.

At that point, the Shadow Lady ordered him to stop. Detective Skaggs fell to the floor of the limousine. Before he could lay down completely, he thought it was over, but nothing could be further from the truth. The Shadow Lady instructed the henchman to drag him by his feet from the limousine.

Then the henchman continues to beat Det. Skaggs within an inch of his life by throwing a series of punches and kicks to his body, as if he was tenderizing a side of beef before cooking. Then the henchman pulls out an 8-inch straight razor and began cutting him in between the kicks and punches. You could hear sounds of his bones breaking, with every thrust of the henchman's force.

The henchman started to tire out, then looks toward the Shadow Lady as she gestures to him to keep going. "Do not stop!" the Shadow Lady replies.

After five more minutes of being broken and beat-up, she looks down at Detective Skaggs. Looking at his broken and bruised body, which had been beaten to a bloody pulp, it looked, to her, like he would see death soon.

The tall dark henchman, almost simultaneously with the Shadow Lady's command, The dark henchman released Detective Skaggs. Detective Skaggs lied there in a puddle of his own blood and looked up into the frigid, rainy night sky. As the dark night consumed yet another victim into its bosom once again.

His thick blackish red blood washed away from his body into the sewer drain, leaving no evidence of foul play, except for the broken bones, cuts, and bruises. There he laid in the streets of the most dangerous neighborhood in the city, Fifth Ward.

The henchman opens the door to the rear of the limo to let the Shadow Lady out. She walks around the back of the limo to where Detective Skaggs is lying. She bends down to Detective Skaggs and whispers into his ear, "I will find the music boxes. I can assure you of that. No thanks to you! I wonder, what else do you know?" They get back into the limo and the limo pulls off.

Detective Skaggs tries to gather himself together to stay warm.

Then a homeless, old, white guy approaches. His walk was staggered, constantly leaning back and forth like he was about to fall somehow; he kept his balance amazingly. He looks down at Detective Skaggs, oblivious to his condition. That is just how intoxicated he was. The old wino says to Detective Skaggs, "Hey, bub, get your own sidewalk. I got here first. I was out drinking with some friends. I have a dog friend, a cat friend, and my rat friends. I have a lot of those just up the street. For some reason, they did not really like each other. I do not understand why. Just like that the party was over."

Det. Skaggs was out cold. The drunkard was rambling to himself, recalling a dream he had when he passed out from drinking too much earlier that day, no different than any other day up the street. With the exception of a badly beaten Det. Skaggs he had briefly as a guest.

Then the drunkard says to an unconscious Det. Skaggs, "Are you listening to me, bub, well, ok." In a slurring, spitting voice the drunkard says, "It's becoming a real problem trying to find good company these days so well."

He proceeded to roll Detective Skaggs off the sidewalk, that he just so happens to be lying right in front of his little cardboard house, saying, "Damn it, brother, you are drunker than me."

The homeless guy stands up, dusts his hands off thinking he had mud on his hand but was thick clots of blood. Soon after the bum staggers back to his cardboard box. He gets in and closes the cardboard door behind him.

CHAPTER 12

The GPS

An hour later Elise, Foxy, and Silva finally reach the apartment. They quickly exit the vehicle in front of the building. Fitts, the doorman, opens the door and the girls quickly walk in. Elise goes to the front desk and asks the attending clerk, "Has anyone been here asking for me?"

The female attendant replies, "No, ma'am, not that I know of and I've been here at the desk all morning."

"Ok, thank you," Elise replies.

"You are welcome; have a nice day, ma'am," the attendant says.

Elise walks away thinking to herself, "If you only knew."

Foxy and Silva hold the elevator door as Elise hurries to join them.

Elise pulls her GPS gadget from her hip to check if Kawanna's signal from her phone is still present inside her condo.

Elise says to the girls, "Well, according to the GPS, she's still here inside my apartment."

Foxy replies in an anxious, gritty tone, "Maybe she came earlier before the morning shift, maybe they didn't feel the need to log her in." Elise says, "Wishful thinking. They log everything, girl." Foxy replies, "Well, let's go see, girl. This damn elevator can't go fast enough!"

Silva in compliance agrees as she jumps up and down, as if she had to pee badly.

Elise notices Silva's up and down motion and says, "Girl, you alright? What is going on with you?"

Silva replies, "I have to pee, like a muthafucka!"

Elise reacts in a surprised and tickled manner and says jokingly, "Aww shit, girl, you gotta take a piss?"

Silva looking back at her barely getting her words out but straight enough to be understood as she holds on to her bladder muscles, which seemed to take all her strength as her bladder weakened even more when the elevator doors opened.

They all seemed anxious when the elevator door finally opened. Then Silva ran out of the elevator first and headed straight for Elise's apartment door. There she stood waiting anxiously for Elise and Foxy to catch up and open the door. Elise picked up her pace in order to let Silva in.

Elise finally gets to the door, unlocks it, and Silva goes in first. Like a bat out of hell, she took off toward the bathroom. In relief, she closes the door behind her and started to take care of her business.

Foxy yells out to Silva, "That's what you get bitch. I told you not to drink all those sodas."

"I know, girl, you know I'm addicted! Oh shit, my stomach is bubbling. Now I got to shit! Foxy, yo muthafuckin' ass jinxed me!" Silva replies.

Elise screaming in a joking manner at Silva, "Girl, close that door before your ass funks up my whole house. Give a sista a courtesy flush, will ya?"

Silva replies sarcastically, "Hey, you, my shit smell like flowers."

Foxy chimes in, "Bitch, I don't know what desperate ass, brown nosing ass man lied to you. If anything yo shit smell like shitty flowers with yo (ebonics) stankin ass!"

Elise and Foxy burst out laughing!

"All right. I'm flushing!" Silva replies.

Elise turns around, then walks toward her bedroom door, hoping to find Kawanna in her bed sleeping. She thinks to herself, laughing and being happy, even just for a moment, it was a very welcoming emotion.

Then she has a flashback going back eight years, all the way back as a senior in high school where she first met Foxy and Silva at a recruitment facility. Elise had signed up for two years in the naval reserves. She had it all figured out. Do her two years in the naval reserves, while at the same time attend college. Everything doesn't always go as planned. She took her lab tests. Her test scores were off the charts. She was already athletic in high school. She played softball, tennis, and won state three times in Division 5A basketball.

A counselor slash recruiter saw her SAT scores as well as the final grades from her honor classes from her high school curriculums. He knew he had a very special candidate in his midst. He picked her to join a, well, let's just say a governmental higher education program and leave it at that.

The recruiter spoke to Elise on several different occasions and finally convinced her to take a leap of faith. This was for a chance of a lifetime.

It took Elise one week to decide. The recruiter thought he convinced her but it was her mother.

For some reason, everything made sense when her mom explained it to her. Then Elise remembered she had *that* conversation with her mother all those years ago. One thing stuck out; it was something she would never forget. Her mother told her, "Do it for you! Recognize the challenge. So that the challenge will have no power over you. Become a part of something bigger than yourself. It will be remembered and appreciated for generations to come."

Elise looks up at her mother, "Ok, Mom. I know recognize my challenge, right."

Elise's mom looked down to her daughter with so much love and joy in her eyes. She says to Elise, "When the time is right, only you will know what your true challenge is, and above all, take care of family."

Young Elise replies, "My challenge is helping those who can't help themselves."

Filled with happiness of joy, Elise's mom says, "Well, my work here is done!"

Elise replies, "You're Mommy, Mom. Your work is never done."

"I know, I know. You know I'm not going to be around forever. I need to know you and your sister can make it in this world. Another thing, when I do pass on to the next life, I don't want no funeral. Cremate me, then sit me on top of the mantel above the fireplace, so I can watch you guys coming and going. That is all I want."

"Momma! Quit talking like that. You will live a long time. Now please, let's switch the subject," replies Elise.

"Well, ok, baby. Just a thought," Elise's mother replies.

Elise replies with irritation in her voice, "Yes Mom, ok. Now stop." Elise still in her flashback. She leaves her mother quietly walking out of the living room. She goes into her room. Getting ready for bed thinking of the important decision she just committed herself to. She falls asleep and rests the whole night thru.

Elise woke up to the smell of bacon frying. She gets herself together, eats breakfast, kisses her mom goodbye, and drops Kawanna off at her school.

Then Elise reported to her appointment at the governmental higher education program (CIA). Looking at her registration paper, the address is in downtown Houston, Texas, finally she makes it to the building. It seems to go up and up as far as the eyes can see. It was a mammoth of a building, as wide as it is tall. The color was a silvery grey that shined in the sunlight like platinum. It was covered in large, tinted windows, which seemed to cover every inch of the building. All the while wondering to herself, "Why have I never noticed seeing this building before?"

She walks into the building looking around for someone to help her. A heavy set red-haired Caucasian woman, in her fifties spoke with a southern belle accent.

She calls toward Elise, "Are you Elise?"

Elise replies, "Yes, ma'am."

She motioned Elise over to her desk. The lady introduced herself, "My name is Mrs. Petit. Ms. Elise, we've been waiting on you."

Elise had a confused look on her face and says, "My class time says 9:30 a.m. if you look on my master copy of my registration form."

"At the bottom it says arrive thirty minutes early before class," says Mrs. Petit.

Elise looks it over and it *was* written so tiny at the bottom, you could barely see it. Elise started to say something but decided to restrain herself.

Mrs. Petit asked Elise to have a seat in the waiting area. She assured Elise that someone would be with her shortly.

Elise takes a seat. Well, the only seat available. She thought to herself, a *waiting area with only three seats. That's crazy.* She sits down between two people.

Elise greets her accompanied strangers. They speak. Then just like that, it became quiet again.

Moments later, a tall, salt-and-pepper haired, clean-shaven, six-foot, five-inches Caucasian man enters the waiting area with three orange folders in his hand. He had a very loud but respectful voice, yells the candidate's name, one by one.

His name was Mr. Green; he is the person that processes new students into the system.

He calls Elise Bishop's name. Elise replies, "I'm Elise." Mr. Green says to Elise, "Could you please stand up and follow me. Also, I need Pharah Nutall and Linda Kay Lopez."

"Here!"

"Here!"

The other two candidates replied eagerly.

Mr. Green looking with excitement and gratitude, lowering his voice, he says to the young ladies in a joking but sincere and honest manner, "Where have all of you been, all my life?"

They all stood up, puzzled, dazed, and confused by his remarks, nervously, without knowing, simultaneously, they all remain silent.

One of the young ladies had a half-smile.

The second had a half uncertain smirk.

And the third had a confused expression, not knowing what Mr. Green was talking about.

Then Mr. Green remembering to himself, saying out loud, "Oh, yea. About what I'm talking about I can already see the expressions on all your faces, please relax. Ladies, I've reviewed over 75 candidates' applications. Well, for now, let's just call it *special schooling, in addition you all will be put in different areas of the armed forces to groom your talents to its fullest potential*."

The girls shake their heads in agreement.

Mr. Green tells the girls exactly what he meant. That they were the *best of the best* and exactly what they were looking for. He says to the girls, "You girls will be here for the next two years. I mean not here, per se, but here with this organization but you can call it *special schooling until we feel you're ready, then you graduate*."

"See what I did with that? You know, *Special Schooling*. Never mind. I know that I have a dry sense of humor. My wife tells me that all the time," says Mr. Green.

Elise quickly speaks up, saying in a half loud voice, while raising her hand, "Sir, sir, I drove myself here. I did not know I would be staying, Mr. Green."

He replies, "That's ok, oh all of your people have been notified, so no worries there. I'm sorry for being vague, but it's protocol. We never indulge sensitive information such as this until you are physically standing in front of us. Do you own a bright green 2000 Honda Civic?"

Elise replies, "Yes," in surprise. "How did you know that? Oh, I know, it's the cameras in the parking lot."

He replies with a straight face, no nonsense of any kind, "The day we were interested in you, that was two months ago, we had this information."

Elise was baffled by his remarks and replies, "So what you're telling me is I was spied on?"

He replies, "We don't see it that way. It's like watching our investment. We do have very smart people working for us. We feel that you would be a nice asset to our family. Now, it's not unusual to have a profile on an individual given the proper resources. You would be amazed at how much information one could find by using a person's social security number."

Mr. Green continues, "No need to be alarmed ladies. I am your mother, father, boyfriend, girlfriend, close friend. Whatever! I'm it!" Then in his next breath, he starts to chuckle, Mr. Green says, "I'm just kidding, you guys; you didn't think that was funny either." The girls looking back in horror speechless, they just shook their heads no not funny from their expressions on their faces. Mr. Green says, "Ok, moving on."

"The only one you trust from this point on are the ones you are looking at right now. Your past as of this moment is DEAD, rigor mortis now fertilizer. Your future begins now; new life will grow into bushels of three beautiful flowers. That can only yield a blossom of loyalties. Gardened by the three of your hands. And your three hands alone."

Then Elise snaps back into the present.

She enters her room; the bed has not been disturbed. Now she is freaking out!

Screaming out loud, "WHERE IN THE FUCK IS MY SISTER!"

Foxy and Silva run into the room and try to calm her down. Silva runs to the kitchen to get a bottle of water out of the refrigerator for Elise.

Foxy sits her down and begs her to calm down.

Foxy says, "I know, baby, I know. We will find her, you know this. I PROMISE! Let it out, let it out. I know it hurts."

Silva slowly sits down next to Elise, gently grabs her attention, and slowly gives her the water. As both girls wrap their arms around Elise, she screams, bawls, sobs constantly crying. Looking as if someone has thrown a bucket of water on her face, but she was drenched with tears.

Foxy signals to Silva to get a wet cold towel for Elise. Swiftly Silva gets up to get a towel from the bathroom. She wets it with cold water from the bathroom sink. Then brings the cold, wet towel back, placing it on Elise's face.

Then Elise holds onto the towel as Silva lets go. Elise continues sobbing and sobbing. Leaning on her only two best friends in the entire world, for moral support.

Asking in a very sad and hurtful voice, rhetorically of course, "Why is this happening to me? First my mother and now my sister is missing. What did I do, GOD! To deserve this, is this a test 'cause right now I don't know what the FUCK TO DO!"

Foxy and Silva both shaking their head in agreement with Elise. They stay at Elise's place as they continue to work, hoping for a miracle to happen; at this point, they were all out of fresh ideas. Perplexed about the chain of events wondering where the GPS signal was coming from, and in that moment, a miracle happens.

Silva hears a beeping sound. She gets up looking for the sound. Then Silva says, "Do you hear that? Or am I tripping?"

Elise and Foxy wait a moment.

Then Foxy says, "I hear something too!"

Elise replies very curiously, "I do too!"

Then they all say at the same time, "Where is it coming from?

They all notice that they said it at the same time. Then they all laugh for a second.

Then Foxy and Elise get up to help Silva hunt for the beeping sound. They wander around and around the apartment. Foxy goes to the kitchen, Elise goes to the living room, and Silva goes to the bathroom. Then they all notice that the sound came from the bedroom. Then the beeping stops.

Foxy runs toward Elise, she whispers, "I don't hear shit anymore, y'all."

Silva comes out of Elise's bedroom motioning Foxy and Elise to come check out what she has found. They both silently comply with Silva, wondering to themselves what the hell is really going on.

Foxy and Elise enter the bedroom. Silva moves out of the way; following close behind them, she points to Elise's lazy boy chair. It seemed like to the naked eye the chair was beeping, because of Elise's unusual daily explosion of clothing thrown on the chair and around the room for that matter. Clothing that did not make the grade for that day. Passed over, picked thru outfits that she chose not to wear on any given workday. There were over twenty-five to thirty outfits thrown on top of the chair and all over the room. It was obviously normal to them, which to me, the narrator, it was very unsettling. Anyway, moving on.

Elise replies to Silva, "Are you sure, Silva?"

"Yes, I am sure. I hear the beeping noise coming from the chair," Silva replies in a whisper.

Elise looks at her and whispers, "Are you sure?"

Silva replies, "HELL YES! I am sure for the (ebonics: meaning multiple times) ump-tee-th time."

Foxy walks toward the chair at the same time telling Elise and Silva, "Move away from the bedroom." Thinking to herself, this could be a fucking bomb. Not to alarm the girls, she keeps her thoughts to herself.

Foxy gently starts removing all of the clothing from Elise's lazy boy chair. Garment by garment. Until there was nothing left but a white square box. The size of the box was big enough to fit enough C-4 explosives to level Elise's whole floor.

Everyone brings themselves to calm down.

Foxy says to Elise, "Do you know about this package?"

"Ahh damn! Yea, Yea, Yea, Yea! I got that in the mail this morning. That is the package I was trying to open before you guys called me to hurry up for work. Remember?" replies Elise.

"I tried to open it this morning, but I kept getting interrupted by my landlord about renovation stuff."

Silva looks at Elise with a puzzled expression on her face.

Elise says to Foxy and Silva, "Never mind." She walks past Foxy and Silva in the living room. She carefully picks up the package. She starts to unravel the package piece by piece as Foxy and Silva look upon Elise in suspense, wondering what in the hell is it!

Elise gets to a brown cardboard box, after getting thru all the paper it was wrapped in. She carefully started to pull up the lid of the package. Suddenly, everybody's phone started ringing at the same time. Elise drops the box in an unexpected scare from the noise coming from their phones.

Foxy and Silva hearing this, caused them to lose their minds in that moment. In a heart stopping scare and without any warning, Elise, Foxy, and Silva instinctively run away from the package, throwing it back on the chair from where they got it. Almost in an instant they all dive to the floor toward the living room. But before they could hit the floor, that is when they knew that it was just the sound of their damn cell phones.

Embarrassed Foxy jumps back up to her feet along with the others. Looking at each other dumbfounded, they started to giggle a little bit.

Silva says out loud, "Serious, that was some crazy shit! We need to get another line of work. This shit here gets more fucked-up by the day."

Elise and Foxy nod their heads in agreement as everyone collects themselves. Almost simultaneously, they begin to giggle and laugh.

They all say at the same time, "This is the best job I ever had." Constantly trying to keep focus always turning horror into sunshine. They all look at each other and say, "That's right, BITCHES!"

Silva says, "Ok, ok. Calm down. It's the Boss at headquarters."

Foxy picks up the package, then Elise says, "That's got to be a sign. Let's hold off on opening that package until we get it checked out at headquarters." Everybody agrees, then checks their phones.

Then Silva quickly says, "We are ordered to come in. Right now!"

Almost in unison the ladies type in their codes. A message is sent back, *confirmed*, to their phones.

They rush to headquarters; carefully retrieving the package in hand, they secure it. Then girls exit the apartment, hoping that the higher-ups could shed light on the mysterious package that has unfolded sort of speak.

CHAPTER 13

The Director

Upon their entrance into the CIA building, as they were getting checked Foxy gave one of the guards the package and told him she wanted a complete report on its contents asap. The guard responded in a salute about-faced and quickly rushed to complete his order. Then two more of the guards escorted the ladies down a hallway that they had never entered before. Then they escorted them into an elevator. The ladies were puzzled because there were no numbers on the panels, but there were alphabets. One guard placed a long silver key inside the panel lock, and the second guard put a golden key inside an opposite lock side panel of the elevator. They turned them simultaneously, then opened up a small silvery square-like miniature door area just above the lock-and-key base.

There, behind the silver square door, was a digital red screen. The guards asked the ladies to stand to the rear of the elevator. Guard 1 bent down close to the floor and pulled a white bar out of the center of the elevator base. Then guard 2 grabbed the tip of the mechanism and pulled it up vertically a few inches more above the floor. It reached four feet from the base of the elevator. There was another metal rod which guard 1 pulled out horizontally from the top of the first vertical rod. The rod extended out two feet.

One by one guard 2 asked the ladies to stand behind the metal rod and be perfectly still. Not knowing what the hell is going on, they all began to ask questions. The guards paid no attention, then the girls all complied with their request.

Then the red screen on the panel sent out a laser that read each one of them from head to toe. As each individual scan was completed, the screen turned blue indicating the process was complete. Guard 2 pressed the alphabet Z on the red screen, and everything, as quickly as it appeared, disappeared.

Then the elevator began to move down and then proceeded to move sideways, which really freaked them out! As they approached their destination, the elevator door opens. The guards step to one side allowing them to exit. Then the elevator door shuts behind and just like that the guards were gone. Elise, Silva, and Foxy were just standing there, then they hear a familiar voice. It is their boss. Section Chief James Williams. Accompanying him was the CIA director, Harold B. O'Hara. Director O'Hara said to the ladies, "I know you have a lot of questions. I prayed that this day would never come. However, due to the latest events, I felt we had no choice."

Elise became perplexed wanting to know exactly what the director was talking about. Speaking out of turn, she anxiously replies back to Director O'Hara's statement. "What!!! ARE YOU SERIOUS?!! I thought we knew everything, Director sir?"

Director O'Hara replies, in a condescending tone, "Please, ladies, this way. I'll answer all questions in a moment." Then motions them to walk with him to a more secure location inside the underground sector, which was secretly hidden.

Foxy thought to herself, there is some sinister shit going on. Then the girls one by one reply back to the director, "Yes sir."

Even though the girls were still starstruck on what they were hearing and what they had seen, they still managed to keep their composure as they all walked inside a white room filled with high-tech computers and monitors. Surrounding the whole room, filled with at least fifty bodies attending them. There were bar codes, maps, and graphs of every sort so extraordinary, one could never imagine.

Then the director asks the ladies to please take their seats, which was separate from the herd of agents. He motions to his personnel to exit the room. Like a squad of synchronized swimmers, simultaneously they converge toward the door. In a matter of seconds, the room was clear. Director O'Hara commences with the briefing. As the ladies wait for an explanation for the reason they were called to headquarters, they were also wondering why, in the hell, they never knew this place ever existed!

CHAPTER 14

The Interrogation

Hours earlier, just a little before midnight coming upon Friday morning, we come back to Detective Skaggs. He tries to wake up in and out of consciousness. Gradually he lifts his head up from the cold, dark streets. He thinks to himself, Man, god loves me. A miracle happens: he finds himself about twenty feet away from a mailbox to the left from where he was lying. He musters every ounce of energy he has in that moment, as he reaches into his socks pulling out a letter with a special code written on the outside, which meant super important rush straight to the CIA Headquarters. Then he crawls to the mailbox. Slowly, pulling his body up toward the mailbox slot, quickly he drops a letter inside, then falls to the ground and unconsciousness overtakes him again.

Then, moments later, he feels a tug on his arms and legs. His eyes barely open. He cannot make out what was going on because of his weakened state. He could feel a thrusting lift, in that moment he felt like air. Powerlessness ran through his mind; if this was his end, so be it. Coming to once again, he noticed he was floating; before he knew it, someone or something is carrying him somewhere! Then he fades out again, his mumbling stops.

Suddenly one side of his face started to alert him that there was a stinging sensation causing him discomfort in that moment he came to. His lungs started burning! It was like he was breathing in water made of fire instead of air. Due to the immense pain, abruptly he is awakened to a hand slap and a bucket of water in his face, as he quickly coughs up the water out of his lungs frantically.

There, staring right at him, was a small four foot, six inches, half-black, half-Korean woman with snow-white hair; she looked like she was about eighty years of age. She was a quiet lady but very aggressive, at least that is what Detective Skaggs assumed as she pulled back with the bucket in her hand. Gasping for air, his face soppy wet, glistening from the downpour of very cold water thrashed upon him.

Then Detective Skaggs heard a very soft gentle Caribbean dialect followed by a younger woman's voice. In a very chauvinistic tone, he thinks to himself, Another crazy, fucked-up in the head ass female speaking out toward him. Then seconds later she appeared. A half-black, half-Korean younger woman. Here is the kicker, Detective Skaggs thinks to himself. She spoke with a Jamaican accent!

As she walked into Detective Skaggs's sight, his eyes were directed downward toward the floor. He was exhausted from the night before due to a fantastic ass-whipping.

He was in tremendous pain! Finding time at all for rest! He gradually lifts his eyes and looks upwards, toward the voice that was speaking to him.

There stood a woman wearing long red leather stiletto boots that was complimented by her upward body attire consisting of red leather shorts and a bikini-like red leather bra. She stood about six feet tall with creamy almond skin. Long very curly jet-black hair. She was very thick in all the right places.

She had a very provocative presence about herself; heart dropping beautiful feminine body every man would instantly lust for. With beautiful big round light-brown eyes. Topped off with cherry red lipstick that seemed to go on and on forever on her full, plump, luscious lips. That would bring most men to a paralyzing stump of obedience. No matter how tall or how deep a man's wall of tolerance was, she would bring them tumbling down. Into her complete submission and utter control if you did not have a trained mind. Unfortunately for them, he did and it showed. The beautiful woman's name is Rachel Fay Chong.

Detective Skaggs speaks in a worn-out voice to the new figure that just invaded his presence. "Who the hell are you supposed to be?" Detective Skaggs asks with a trembling weak voice.

Rachel replies, "I can be your paradise, or I can be your hell. It's your call, blood-clot!"

Detective Skaggs replies, again very weakly, "Who the fuck are you and what did that bitch do to me speak English?" referring to the old lady.

Rachel replies, "My family, Lady Lou. You're right, she does not speak English."

Detective Skaggs knew right away this was an interrogation. Being a seasoned vet in counterterrorism, he knew automatically. So, he purposely tried to get a rise out of the two women in order to see if they would reveal any vital information. Basically, he wanted to know if they were idiots. At this time, unfortunately for him, he would get his reply tenfold. Waiting for a response hoping to get any information that could be pertinent to his case.

He looked over at Lady Lou and says to Rachel, "She is so short; she reminds me of a snow monkey. Is she your pet?"

Rachel replies, "You shouldn't have said that."

Detective Skaggs screams in pain as Lady Lou sliced his right arm wide open, but not life-threatening, with a very sharp surgical blade. This all happened before Rachel could even finish her statement to Det. Skaggs. Then Lady Lou giggled and heckled briefly as she became overwhelmed with joy by Det. Skaggs's reaction; immediately, she wrapped his arm up so tight and snug before he could bleed out. Then Lady Lou stepped back and smiled as she looked back at Skaggs liked nothing ever happened.

Detective Skaggs's anger-filled voice cut thru the silence like a speeding bullet hitting its mark. He loudly screams saying, "I THOUGHT YOU SAID THAT CRAZY BITCH DIDN'T SPEAK ENGLISH!"

Rachel replies, "I said she did not speak English. I did not say she did not understand English. Her tongue was cut out of her mouth by her own hand. She took a vow of silence for the rest of her life. She did this for her religious purposes, and for inner peace, and a whole lot of other ancient shit I care not to get into too.

Detective Skaggs started to say more negative comments, but Lady Lou looked directly at him like an angry lioness waiting to pounce. He realized it and kept his mouth shut but not before he gets off a whisper saying, "That was too easy anyway."

Rachel says rhetorically, "Do you think you just happened to be on my doorstep by chance? No sir, my friend. You were purposely left here. Well, I hate to be the bearer of bad news, but the Shadow Lady knows you've been compromised, as I'm sure you already know. The Lady, the Shadow Lady I might add, has tasked me with just how much you've been compromised. So that being said as of this moment your ass belongs to me, Detective Skaggs." Det. Skaggs started to laugh. Then suddenly Rachel loses her temper by yelling and screaming at the detective in a crazy psychotic way, hoping to scare information out of him.

Detective Skaggs jumped up trying to break free from the chair and ropes that he was securely bonded to, in a desperate attempt to physically harm Rachel. So frustrated and angry he suddenly stopped, it was like out of nowhere he became lucid, and again he tries to make light of the situation by jokingly jumping up and down over and over again, then he says sarcastically, "Oh please, master, please, please, don't kill me!" Then he gets serious as he straightens up, he continued by saying, "Bitch, I am not saying shit to you, the Shadow Lady, or that Lil' Chinese Wookiee-looking bitch standing next to you!" Then he sits back down.

Rachel looks at Detective Skaggs and laughs at his little quick-witted jokes and his little Oscar winning performance of being a typical man. She thought to herself, I am not surprised. She holds Lady Lou back from Detective Skaggs. Then she slowly walks around and around Detective Skaggs as she slowly takes her hands and arms and began to caress his face, down to his chest. Then she straddles Det. Skaggs in the chair, by jumping up high in the air. She flops down hard on his lap, catching him completely off guard. Rachel started to kiss his neck, while one hand was firmly around his neck. In the other hand, out of nowhere a razor blade is caressing Det.

Skaggs's throat. That came from the tip of her tongue so strategically, as if her tongue had a thumb and an index finger. Out of her mouth! She grabs hold of the blade and slowly started gently rubbing the blade in a circular motion on his neck. This caused Detective Skaggs to sit perfectly still as a bead of sweat runs down the side of his right cheek. Racheal replies, "What was all that sweety you were saying earlier? Come on, talk to me now, baby. I want to feel your voice on my tongue." Rachel started licking Det. Skaggs's Adam's apple like a cherry on top of a banana split. Det. Skaggs's response utter silence. Rachel says, "Yeah, that's what I thought."

Then Rachel's phone starts to ring. Then Det. Skaggs thinks to himself at the same time shouting out loud saying, "I know it's that Shadow bitch on the line."

All he could hear was Rachel's response to the phone conversation.

One second she said, "The muthafucka was nearly dead! Lady Lou had to give him a blood transfusion to keep him alive the first two hours, not to mention wrapping broken bones and flesh gashes throughout his whole body. Then he has been coming in and out of consciousness! We gave him an IV to hydrate him, so that the formula can work more efficiently. The truth serum you provided has another ten minutes to work."

Then a long pause went by, Racheal starts to scream in terror saying, "There was not enough time. I'm not there yet. Please don't, I beg you!"

Suddenly as he looked at Rachel, her whole facial expression changed to a worry and disbelief. Then after a long pause, Rachel replies, "We had a deal, you lying double-crossing bitch!" She, gasps for air trying to calm herself down. Then she began to say, "Hello!!! HELLO!!!" screaming in a rage. She threw the phone on the floor, breaking it in a million pieces!

Detective Skaggs felt no sympathy for her at all, all the while he was still trying to get loose. Then he saw Lady Lou fall down to her knees bawling and sobbing uncontrollably.

Rachel puts a blindfold over his eyes. Then that's when Detective Skaggs was very, very worried! He knew 100 percent that the only person that could leave that kind of impression of pain and misery in their wake was none other than Momma King, a.k.a. the Shadow Lady. Then moments later not long after the little disturbing phone call, Detective Skaggs heard footsteps, tapping on the worn-out wooden floor. As he sits there, the sound was faint at first. Then the footsteps got louder and louder, as every step was approaching them quickly. Det. Skaggs not being able to see clearly through the blind fold. He caught a glimpse of a dark figure; his head was cloaked in a thick dark cloth, he could hear a voice spring out into the dry creepy, heart-wrenching silence. He recognized the voice. In relief, he noticed that it was Rachel. Scared shit less, not being able to control his fate, nearly drove him bonkers. Hoping to himself he could get loose before another surprise visit from the black, dark henchman of the Shadow Lady!

Even though he was most likely to die, in these last moments what mattered most deep down inside, he rather have her, then the dark henchman of the Shadow Lady. She was the lesser of two evils.

Rachel looked down at Detective Skaggs, then pulled his blindfold off. Detective Skaggs looked around frantically in horror, waiting on a bullet or a knife to slit his throat. But it never came. As Detective Skaggs focused in order to see who was there in front of his face, standing there with a blank look on her face was Rachel.

Detective Skaggs yelled out loud as he focused his eyes on Lady Lou. She was angry and severely frustrated and blamed Skaggs. She reached out by slapping the shit out of Skaggs, again! He went out and came back to a few seconds later saying, "What just happened?"

As Detective Skaggs becomes completely focused again, he looks around at Rachel then again saying, "Can you please keep this scary, mean bitch away from me?"

Rachel replies, "Settle down you two! We have bigger problems; about thirteen to fifteen of them coming up the stairwell, from the sound of the shuffle of feet coming this way. It's Momma King's cohorts are coming to end us."

"You need to untie me right now or we're all dead," replied Detective Skaggs.

"In case you forgot, you were caught as a spy in her organization! I failed to obtain the information in time. In turn, that bitch just killed my uncle! My aunt's husband!" replied Rachel.

Before Detective Skaggs could reply to Rachel, there was a knock at the door.

Then there was a voice screaming out loud! "This is the police! Come out with your hands up!" In all the confusion, Racheal started to untie Skaggs.

Lady Lou looked through the eye hole of the door. Detective Skaggs slowly but quickly gathered himself and stood to his feet.

Rachel screamed at Lady Lou, "DO NOT LOOK THROUGH THE DOOR!"

Before she could finish her statement, the frail wooden door, in an instant, exploded into a puff of splinters and white smoke. In horror, at the same time, they watched the top part of Lady Lou's body explode into a bloody mist, with human flesh fragments flying all over the place. Then Lady Lou's lower body spun around from the force of the blast. Her nerves were still active for a moment, long enough to run and drop down into Det. Skaggs's lap. A piece of her went into his mouth; he lost his mind. He started throwing up while ducking bullets all at same time. Thinking to himself, And the hits just keep on coming. Then his training kicks in, and he becomes calm and focused. In complete contrast to Skaggs, Rachel had a different approach as she yelled in horror! Not for herself, for the Shadow Lady's cohorts, SO LOUD and very, very sure of herself she says, "I AM GOING TO KILL EVERY LAST ONE OF YOU SONS OF BITCHES." Then there was a painful cry as she screamed in seemed like unbearable pain, but there was no time to mourn in that moment so kept repeating these words a lot saying, "THAT BITCH! SHE KILLED LOU! SHE Killed MY LOU! She KILLED LOU!" Over and over again, as if in disbelief. Then after emptying her fifty rounds to five already loaded riot shot guns. Considering her career path was not a surprise that she had them mounted on her walls throughout her apartment. Laying waste to everywhere and thing that moved. For three minutes, no one could get close enough to her for a clear shot.

Still in shock and horror, she managed to grab her 2 Glock 9 mm pistols from the kitchen table, then started to unload at the door. With a series of automatic bullets that seemed to rain from her hands. Like a hell-bent electrical rainstorm, with no end in sight.

Just yellow and red lights flying inside the room, across the room, plunging into the walls! Taking large chunks out of the walls! The bullets were destroying everything in their path! Totally destroying her apartment.

Rachel drilled the first assailant in the head! Instantly his head exploded from his neck! Two more hooded assailants followed behind the first. Pouring inside her apartment! They immediately separated, hiding behind turned over furniture inside the living room waiting for their time to return fire.

Det. Skaggs falls down to the floor inside the kitchen! As he crawls toward the refrigerator, he pushes it down horizontally on its side! While he is constantly ducking and using it as cover from the array of bullets that seemed to be coming from all directions! Then he hears a loud thumping sound from upstairs! Then another thump! That was coming from the kitchen window, where the fire escape was located.

He thought to himself, There were two more assassins in their midst! At that moment he knew where the threat was coming from! Getting a glimpse Det. Skaggs noticed, they were all dressed like ninjas, all in black with red hoods on their heads. This attack seemed thought out. He came to one thought that it was none other, the Red Scorpion Clan. They were a band of mercenaries, ran by a sinister lunatic whose name is Darius Khan, a.k.a. the Dark Stranger!

He is Momma King's top assassin. He does mostly her dirty work. He stands about six feet tall and is of Japanese descent. He has a very dark red like tattoo scar across his face with long jet-black hair. His left eye is ocean blue; his right eye he keeps covered with a dark black patch. He wears a half-faced mask that covers his mouth! His weapon of choice is two, twin silver blades. Attached to one end is a custom-made very strong, thin, red steel chain that extends outward of five feet from his hands; nothing is safe within his grasp. The chain seems to be connected to his wrist. The chain retracts backwards and forwards to a custom-made cartridge of some sort beneath his robe, attached to his right arm! He is extremely dangerous and very skilled with this device, not to mention a master in every martial art known to man.

Detective Skaggs quickly snaps from his thoughts as his attention was completely overtaken by Rachel!

She is screaming at him, "GET UP!" as she continued to fire at the henchmen! She stops for a moment in order to reload! That was all the time the henchmen needed to start another melee of return gunfire! Rachel knew they were outgunned! Det. Skaggs did not even have a weapon. She quickly helps Detective Skaggs to his feet.

While he is limping on one leg, screaming in agony because of the additional wounds that were inflicted by the late Lady Lou.

After reloading, she grabbed the half-drugged Detective Skaggs and pulled him into the laundry room, which is located next to the kitchen! As they try to flee, just past the kitchen stove down the hallway, Rachel continued firing at the intruders. Keeping them distracted long enough to come up with a plan. In between all of the exchanged gunfire, Det. Skaggs constantly keeps his head down, cannot believe they are not dead yet. Then Rachel quickly opened the dryer door in the laundry room. There sitting inside the dryer was a home-made bomb with a timer. She sets the bomb to explode in ninety seconds!

She helps or rather yanks Detective Skaggs to his feet and quickly moves the washing machine away from the wall! There behind the washing machine was a steel door in the floor. She kicked the door handle open, it sprung up! There inside was a yellow, long, wide, round tube. She pushes Detective Skaggs inside the tube, then grabs the handle as she pulls the door hatch down! The intruders broke through the kitchen, saw Rachel jumping inside the hole, as she continued firing a hail of bullets! She gets hit in the arm right before she closes the hatch. A wounded Racheal still locks the hatch, then follows closely behind Skaggs.

The intruders tried vigorously to open the hatch lock! Three more men made it to the kitchen, they were wondering what the hell was going on! Then one heard a ticking noise, then opened the door to the dryer unit. There staring right at him was the ticking sound of a bomb, which was counting down from five seconds! He tried to warn the others of the bomb as he took off running out of the kitchen! Right past them. He made it to the living room before the bomb went off! Instantly incinerating the intruders along with everything inside the apartment, while leveling three-fourth of the apartment building!

CHAPTER 15

KLOL CHANNEL 10 NEWS

Luckily, the apartment building was abandoned. For the most part, it was only occupied by Rachel and her Aunt Lou! There were no innocent bystanders harmed in the explosion! Twenty minutes after the blast, fire trucks, ambulances, and a news crew were on the scene.

KLOL Channel 10 news reporter named Tim Berry. He is a short dark, black man, with a muscular build, dark brown eyes, with salt-and-pepper colored hair resembling a white cutout version of a MSNBC anchor Brian William. He stands about five feet, ten inches and weighs 175 pounds. He graduated from Howard University. He is a very ambitious type individual that everyone at work hated because he had no sense of humor and was a typical royal pain in the ass; even though they hated to admit it, they knew he was really great at his job.

Tim Berry showed up at the scene with his cameraman and his mic in hand; without a moment to waste, quickly they find the best spot and get set up, just before Tim started to report the news.

The cameraman looks over at Tim, the cameraman holds up his hand and counts down 5, 4, 3, 2 … points at Tim. Tim Berry begins saying, "*The Blast in the Ghetto*." Reporting Live in the Fifth Ward. According to the fire marshal, there was a bomb that blew up an apartment building called the Navarro, a building that was built in the early 1960s. That's been nearly condemned for the last twenty-five years. According to reports, it's been home to only one family for the last five years. At this moment that is all we have. The Navarro is located at the corner of English and Old Spanish Trail Street. The bomb blew up between the third and the fifth floors. Hold on, Hold on, I'm getting more information right now …"

He puts his head down and presses his hand to his earphones to block out other noise pollution that was keeping him from hearing this vital information.

Looking back at the camera he continues, "I've just been informed that there are six bodies that have just been discovered inside the rubble of the building. Five males and one female. However, we do have some good news. My sources tell me there was one survivor. He is a homeless man, who was buried under the rubble. He was rescued by the fire department shortly after the explosion. They dug him out of the rubble. They say if it hadn't been for layers upon layers of cardboard boxes stacked up on top of each other, along with a bunch of old clothing that he used to keep himself warm, he would not have survived the blast. This is Tim Berry, reporting live from KLOL Channel 10 News."

CHAPTER 16

The Sewer

Minutes later, a very badly injured Detective Skaggs and a distraught Rachel finish their descent from the yellow tube that Racheal had thrown Det. Skaggs in a last ditch effort to save their lives. They both end up at the bottom of the sewer below the city streets. As they quickly try to recover from the appalling, smelly, shitty water that they found themselves bathing in, a surprised and very aggravated Detective Skaggs realized what he had just landed in. It caused him to erupt in an uncontrollable rage; instantly without thinking, he became a shit talking, crazy person. He would not shut up until everything in him was satisfied due to the current events.

He started to yell and scream in disgust mainly toward Rachel saying, "WHERE THE FUCK DO YOU HAVE ME NOW, BITCH! This place smells like hell's fat, funky draws! YOU DAMN STUPID BITCH! SHIT, you should have left me up there to blow up! Instead, you bring me down here with my open fucking wounds just to let me die slow from urine and feces infection! And YEAH, WHO IN THE HELL puts a fucking bomb in a dryer? You are not quite right in the head! Even before that your crazy-ass aunt tried to make chop beef out of me. God bless her soul!"

Rachel looked back at Detective Skaggs waiting for him to finish his crybaby raging tantrum. Then she replied, "Ok, you finished now, Detective Skaggs?" Looking surprised that she could even ask him that shit.

"Why, you thought of some new shit! I can say about this fucked-up day that I might have forgot about!" replies Detective Skaggs.

Rachel replies, "You know, Detective Skaggs, I was trying to save my uncle's life. I failed to do that, and now my aunt is gone. I have no more family. It was not personal only business."

Detective Skaggs quickly interrupts her saying, "Well, muthafucka! It was personal to my ass! Getting choked, burned, stuck, cut, and drowned didn't feel like business to me. Hit in the face, starved, and thirsty to near death! On top of that, given a truth serum that did not even pass the FDA as being safe to use on foreign enemies! It was ruled inhumane and 75 percent of the test subjects suffered massive heart attacks, and soon after, they died Miss Racheal!" Racheal replies,

"Well, you should be thanking God that you were one of the twenty-five percentile, RIGHT? Because I am not going to apologize for trying to save my family, so that means you have another purpose to serve in your life. That is the way I would have looked at it!" replies Rachel.

Detective Skaggs mad as hell replies, "YEA, that would be how your ass would look at it! 'Cause your ass wasn't where my ass was, was it?"

"Well, you rather I'd left you up there to get blown to bits! MY aunt was just killed! Have some FUCKING RESPECT!" replies a pissed off Rachel!

"Well, excuse me if I do not have any love lost for your crazy, deranged, bitch of an aunt that tried to off me!" replied Detective Skaggs.

Rachel started crying, saying in a sincere voice, "I'm sorry! I am truly sorry! Is that what you want me to say? They had my uncle. I had no choice! DO YOU HEAR ME? NO CHOICE!"

Detective Skaggs stopped talking shit. He realizes he might have gone a little too far!

Also, thinking to himself, "Shit, if she says fuck it, I might get left behind to be eaten by these sewer rats, and these are some big fucking rats down here!"

So Detective Skaggs changes his tone a tad bit saying softly, "I'm sorry for your family, but we have to get moving. If you help me, I will help you to get revenge on Momma King."

Rachel stands there sniffling, trying to dry her tears with her sleeves. Whispering under her breath, she replies, "Ok."

Detective Skaggs looks over to Rachel and says, "Now, where in the fuck are we and how do we get out?"

"We keep going north, this way," she replied.

Detective Skaggs uttered, "How in the hell do you know which way is north?"

Rachel relieved that Detective Skaggs changed his angered, bitter attitude for the better. Now she could concentrate on the matter at hand. With a less bitter voice, Rachel replies, "Well, I know you probably think I'm crazy for telling you this but I know it."

Detective Skaggs replied, "Probably. Really."

Rachel looks back at him with an expression like, see, I told you so!

Detective Skaggs catches himself and apologizes again saying, "Ok, how do you know?" He allows her to continue her conversation.

Rachel continues, "Anyway like I was saying, when I was a child, I had childhood friends …"

Detective Skaggs blurted out quickly, jokingly, "You had friends as a child, you were a child?" Then he hurried to say, "Ok, Ok, OK. I am sorry seriously. I just had to say that. I am just teasing kinda. Shit, you know I am due that. You know and I know that I should be choking your ass out and trying to find my way out this fucken place, but I know I can't physically do that so I need you and you need me. So let's cut the tension, agreed!"

Rachel agreed.

Detective Skaggs continued, "Go ahead. Finish your story. I won't interrupt you again."

Rachel looks at him like yeah right.

Detective Skaggs genuinely seemed sincere, so Rachel continues her story. She replies, "I used to come down here with childhood friends. So, I know these sewers like the back of my hand. This way is north, about 500 yards to Clemons Street. Then we turn right at Black Waterfall Street."

Detective Skaggs curiously says, "What the hell is Black Waterfall?"

Rachel started to laugh, "Oh yea, I forget who I'm with. Yes about that, it's a name friends came up with when we used to storm these sewer floors. The Black Waterfall was just a name we came up with when we were playing Mystical Dungeons and Dragons games. You do remember Dungeons and Dragons as a kid, don't you?"

In an aggravated voice, Detective Skaggs replied loudly, "Yea, but that wouldn't have been my guess to why you called it Black Waterfall, you know because of the dirty shitty funky ass water, but yes to your question. I do know Dungeons and Dragons. I'm not stupid." Sarcastically he continues by saying, "Guess what else, I have even seen the movie."

"You know it was a book and a game first, right?" replied Rachel.

Detective Skaggs looks at her with an irritated expression on his face and says, "Yeah, whatever." In a not-so-sure remark. He continued by saying, "Are we seriously talking about this, right now? Get me the hell out of here! Right now, please!"

Rachel purposely took his attention away from the painful walk he was experiencing, by occupying him with distracting stories. Then she says, "We're here, Detective."

Detective Skaggs in disbelief says, "What?"

"We're here. That is the exit, 30 yards ahead. You stay here; I'll go up to see if the coast is clear," replies Rachel.

Rachel pushes the sewer steel plate to the side from up top. Peeks her head out, then she looks back down toward Detective Skaggs saying, "The coast is clear." Then she climbs back down the ladder quickly. Then she gets Detective Skaggs prepared to climb up the ladder in order to exit the sewer.

Detective Skaggs says to Rachel, "When I get to the top, you not going to push me back down to my death, are you?"

Rachel looked back at him with a surprised expression on her face, like *I cannot believe this fool just asked me that.*

Before she could speak, Detective Skaggs quickly blurted out, "I'm just playing! Seriously, You wouldn't, right?" Racheal replies, "Skaggs, if

I wanted to kill your ass, I would have left your ass in the apartment. Think about it. Like you said, we need each other. I just hope you're a man of your word. Now come on shit let us get out of here, Skaggs."

Rachel grabs him and stands him up on his feet, putting him in position. Detective Skaggs takes hold of the ladder. Rachel pushes him upward toward the surfaces to the street level. Detective Skaggs slowly crawls out of the sewer, twisted over and over until his body was completely out of the hole. Then he slowly stood to his feet. Rachel soon climbs out behind him. She saw Detective Skaggs struggling to stand on to his feet.

CHAPTER 17

The Pickup

Sometime later, Friday morning, in the midst of Detective Skaggs and Rachel emerging from the sewer, there standing across the street, near the crosswalk, were two young black African men waiting on the bus. They were two exchange students from Ghana, Africa, there attending the University of Houston. They were in shock as they saw Detective Skaggs and Rachel rise out of the sewers to the streets.

The first African man looks back at his friend and says, with an African accent, but in English, "Did you just see that?"

"I am looking just like you!" replies the second African exchange student.

"I don't believe it! These Americans have cities underground as well! This is truly like everyone says. This place is truly the greatest country in the world!" replies the first African.

His friend replies, "Yes, I do agree."

The first African looks back at his friend nodding his head in agreement again.

As Detective Skaggs and Rachel escape the sewer, they head west down Main Street. Detective Skaggs quickly says to Rachel, "We need to find a pay phone as soon as possible." Racheal replies, "I know there's a coffee shop on Main and Kirkland Street just across the street from the Subway sandwich shop."

"Oh yea. I think I know where I am now. I know a billiard hall about five minutes from here, where we can lay low until help arrives," replied Detective Skaggs.

"What do you mean until help arrives?" replies Rachel.

"I have a homing device implanted inside me to alert the CIA. They can locate my position," replied Detective Skaggs.

Rachel says, "Oh, about that. Lady Lou cut that out of you when you were unconscious."

"DON'T YOU THINK THAT WOULD HAVE BEEN A LITTLE IMPORTANT TO HAVE TOLD ME BEFOREHAND, WHILE WE WERE DOWN IN THE HOLE?" yells Detective Skaggs.

Rachel replies, "At the time I was not thinking about that. I was more concerned about your health and keeping YOU BREATHING."

Sarcastically Detective Skaggs replies, "Yeah, whatever. Damn it! Well, now I guess we'll have to go with your plan, for now."

Rachel looks at Detective Skaggs and says, "Well, let's go."

They arrive at the coffee shop fifteen minutes later. Rachel goes inside first, while Detective Skaggs stayed outside. She checks out everything while talking to someone inside the shop. Two minutes later, the man she was talking to happened to be the manager and the owner at the shop.

Detective Skaggs knew that she knew the owner very well, because soon after their conversation, he kicked everyone out of the coffee shop saying that there was a gas leak and he didn't want to get anybody sick!

The few customers he had left immediately. There were only three people. A Spanish lady and a retired old couple that lived around the corner that came there every day, because it was the only entertainment they had, which was close by.

As they departed the coffee shop, Rachel walked back outside, gestured to Detective Skaggs, and said, "Come on in. It's safe; he's cool," referring to the gentleman inside the coffee shop she was conversing with.

The owner's name was Robert Sparks. He was a red-headed, heavily freckled face Irish man, about twenty-four years old. He was a computer geek, whom Rachel had used in the past on previous jobs.

Detective Skaggs noticed Robert had a little crush on Rachel; you could tell by the way he looked at her. Det. Skaggs jokingly says, "He fancy you girl, you know that." Racheal replies, "Yes, I know. He's been having a crush on me since he was ten years old. I used to babysit Robert when I was a teenager in high school attending Yates High School. His family knew my family because our fathers used to work together at the Budweiser Plant, off I-10 E. Remember that explosion they had years back?" Det. Skaggs says, "Vaguely but yes." Racheal replies, "Well, our fathers and mothers were at work that day; they all died. The tragedy brought us closer like an unconditional love bond—it's complicated. I've been around my aunt and uncle since. My family and the Sparks family worked together for over ten years while becoming very close. Robert's parents needed someone to watch their young son so that they could go to work. I was fifteen years old at the time. I was the only person they trusted to be in their home. Robert was eleven years old. One day when I was babysitting Robert, I let him see my breasts. Since that day, he has been my slave sort of speak. Det. Skaggs says, "Damn girl, a little too much information. Ok, I get it he was your flunky." Racheal replies, "Well, you asked." Det. Skaggs says, "Sorry I did. Come on, let's get inside."

Detective Skaggs walks into the coffee shop. Followed by Racheal. Rachel introduces him to Robert.

Robert says, "Hello, sir. Damn, you look like shit and smell like it too; you too, Racheal!" Noticing to himself that he actually said that out loud, Robert tries to fix what he had just said. Quickly Robert says, "I'm sorry. I did not mean for it to come out like that." In an exhausted tone, Det. Skaggs replies, "I understand. Really, I understand! It's been a long night, Mr. Sparks." Robert says, "From the looks of it, yeah, I'm inclined to agree with you. You need a doctor like yesterday, sir, and please call me Robert." As he tries to shake Skaggs hand but decides not to mainly because, to Robert Det. Skaggs looked to be in a lot of pain, not to mention very filthy.

Rachel quickly intervenes saying to Robert while pushing Detective Skaggs in the direction of the back office, in order to clean him up and let him make his phone call to the CIA Headquarters, "We need to make a call."

Robert, still in awe from Detective Skaggs remarks. Watching him in disbelief, he really wanted to understand what the hell he was trying to tell him!

Rachel goes back to Robert, and as quickly as Detective Skaggs had escalated the situation, she, in turn, quickly deflated it. She saw Roberts's nervousness.

Robert says, "I think Mr. Skaggs is a little upset. I wonder what has him so aggravated."

Then Rachel replies, "Believe me. You do not want to know. You have no idea!"

Robert quickly replies to Rachel, "You know what, you might be right. I don't want to know!"

Rachel and Robert stand outside the room waiting on Detective Skaggs.

Soon after Detective Skaggs makes his phone call, there were agents on the scene to pick them up ten minutes later. Six black SUV's pulled up right in front of the coffee shop. Section Chief James Williams jumps out of the third vehicle with his men dressed in all black military attire from head to toe armed to the tee, like that they all converge in the coffee shop. There were four men standing outside and there were eight men inside looking and searching for any and all threats. When the area was confirmed clear, one of the agents signaled to one of the outside agents. Then Section Chief Williams walked in.

Section Chief Williams greets Detective Skaggs with a sigh of relief and a strong hug. The chief says to the medics that were part of his entourage, "Fix him up real good; he's one of my best men." They comply, then leave the coffee shop, then one of the agents asks Rachel to sit five tables down from where Detective Skaggs and the Section Chief were sitting.

She agreed with no hesitation, blurting out to Detective Skaggs, "Remember what I did for you, Detective."

Detective Skaggs looks back at her in disgust. Thinking to himself, *SHUT THE FUCK UP, BITCH! I KNOW!*

Section Chief Williams looked in surprise like, *what is going on? Who is she?* "Sir, it's a long story," replies Detective Skaggs.

"Well, I have time. Explain," replies Chief Williams.

Detective Skaggs relates the entire story. From the Shadow Lady to the interrogation to the sewers and then to the coffee shop. Then ended it with, "I promised her asylum from prosecution for her full cooperation."

Chief Williams complies with what Detective Skaggs had promised Rachel.

So, the group gathers themselves together.

Rachel thanks Robert for his help and gives him a hug and a kiss on the cheek.

Detective Skaggs watching Rachel and Robert and says, "Are you too finished? Rachel, let's go!"

Rachel thanks and says goodbye to Robert.

Detective Skaggs walks over to Robert, gives him a handshake, and thanks him for his help, then exits the coffee shop.

Robert was still in shock from what he had just seen. His body went limp, then stiff, like he was having an out-of-body experience, as his hand went up moving from side to side waving goodbye. Then everyone was gone in an instant. Thinking to himself what just happened, Robert walks outside and watches the last of the black SUV's quickly disappear, out of sight, as if the event never occurred. Robert thinking to himself that if he valued his freedom, he needed to believe that day never ever happened.

Inside the SUV, Section Chief Williams calls ahead to inform Director O'Hara's assistant, Rayford Chiles, to have the nanotech surgeons prepped and ready for Agent 888, a.k.a. Det. Skaggs. The lion is in the den," says the chief.

Rayford Chiles, in turn, informs the director directly.

The SUV convoy finally arrives at the CIA Headquarters. Director O'Hara is formally briefed of the events according to Detective Skaggs and his new asset, Rachel. He is informed about the Shadow Lady and her operation.

Director O'Hara orders Section Chief Williams to call in the girls. Better known as Elise, Foxy, and Silva. A critical briefing is needed. Chief Williams complies.

Director O'Hara says to Chief Williams, "It's time to tell them the whole story."

Section Chief Williams replies, "Sir, do you think that's wise?" Director says, "I don't see any other way. I feel it's time, Chief. They need to know! These girls are the key to winning this battle between us and the Shadow Lady. We both know what Momma King, a.k.a. the Shadow Lady, is capable of doing." Chief replies, "I sure hope you are right, sir. Because if we are wrong, the world as we know it will be over!"

"Let's hope it doesn't come to that, Chief. God knows, I do, they might be our only chance of survival," says Director O'Hara.

Chief Williams replies, "Agent 888 will be ready for briefing within the hour after his refresh procedure, along with the girls. They'll be here shortly." Director O'Hara says, "Bring them in when they arrive and are ready. I'll be in Chamber Z."

Chief Williams complies and goes back to his office. The director disappears into a secret corridor.

CHAPTER 18

Mission Failed

Meanwhile, inside the Shadow Lady's lair, a.k.a. Momma King. There she sits inside her dark office. It was lit up by red glowing light. As if she was sensitive to white lights. She had a lot of dead stuffed animals, which seemed to take up almost every inch of her office wall. She had a 2000 gal. saltwater fish tank, which was filled with Piranhas. She fed them raw cow flesh and sometimes human.

That is the type of woman she was by nature, ruthless. Just as bad as any man, some would think even worst.

The CIA has been after her for years and they always hit a dead end every time. They have never been able to locate her or the Shadow organization. She always for some reason stays two steps ahead of them. She has as many agents as the governments that chase her. No one truly knows her sinister agenda but her, that being the real reason she's feared.

Momma King is a six-foot, three-inches tall, highly intelligent, not to mention a beautiful dark-skinned, black woman about sixty to seventy years of age, according to some records that are not officially confirmed, files that are not for the public for obvious reasons. Due to her exposure to white energy, somehow has allowed her aging process to be slowed down dramatically. Oh yeah, we will talk about that a little bit later on. She looks like she's still in her twenties. Really, her age can't be determined by just looking at her; officially her birth can't really be proven for now. Momma

King made sure of that. She wears vibrant colorful dresses, with beautiful African art woven into the fabric. It seems like she has a different dress for every day of the year. Like Dean Dornn. She never wore anything twice; she has a thing for fashion like any other typical woman, especially when it came to the latest fashion, she was always on point.

Her shoes, jewelry, perfume, hair, and even her makeup always a picture perfect match. She has very long, beautiful hair that she mostly wears in a bun, with exotic flowers from all around the world that sticks out of her head like a crown. She has large, beautiful white teeth. Beautiful when she smiles. She seemed like an average or normal woman; she looks like she is very loving and nurturing at first glance, but nothing could be further from the truth. The color of her eyes is like two burning suns, flying side by side in the cold darkness of space. A very athletic individual. She loves to swim, run, and lift weights. She is always doing something physical in order to keep herself emotionally fit and mentally sharp.

Except when she is being very creative in killing someone or something for their insolence. That is her true passion some think, also her ultimate pleasure.

The dossier the CIA had on the Shadow Lady, somehow became erased or misplaced; no one really knows exactly. She is completely off the grid. Some seem to believe she killed everyone that knew her true identity. Which includes all of her family and the ones who were not family that knew, which were a very few. The ones that do know are only on three fingers; they worship her, loyal only to her, they would even die for her. One of the three is her. Let's get back to our story.

There she sits in her dark office. Darius Khan walks into the room. Momma King looks up toward him. She quickly speaks, "I hope we have good news."

Darius Khan puts his head down, staring into the floor. Then he says, "The five men that I sent are ALL DEAD. Rachel's aunt was among the dead as well. But Rachel and Detective Skaggs's bodies were not found in the rubble. That tells me they are still alive."

Momma King grabs hold of the corners of her chair! Squeezing them so hard, you can hear the wood cracking! She was still sitting.

But when Khan heard the crackling of the chair, he slowly, gently, took two steps back!

Then Momma King looked up in disgust and said, "Well, that serves them right! It is best that they had NOT returned without completing their mission! I would have killed them with my bare hands then, fed them to my little darlings."

Khan's nervous voice chimes out, "Yes, mistress. I know. I only ask, what do you need of me?"

"Come closer my darling Khan," replies Momma King.

Reluctantly Khan moves closer.

She jumps up from her chair and slaps his face. Then says to Khan, "When I send you to complete a mission, I expect you to do just that. Do you understand?"

Khan fell to his knees. Begging for mercy. Knowing in his heart of hearts, she could or would kill him at any time. The only thing he thought of was not failing Momma King again; oh, also being fish food, that's a given.

Momma King tells him to rise from the floor. Then hugs him and kisses him on the cheek. Then warns him not to disappoint her again.

"Never again will I fail. Unless death claims me!" Khan replies. Momma King says,

"Apology accepted. Now we have to find those last two music boxes. Without them, the plan cannot proceed. So, find me those boxes, Khan!"

Khan replies, "I have men watching the building. We should know something soon."

Momma King says, "That's good. Let me know what transpires and check in on our little guest. See if she is talking yet."

"Ok, mistress," replied Khan. Then he exits the room.

CHAPTER 19

The Truth

Meanwhile, back at the CIA Headquarters. Chief Williams presses his intercom, alerts his assistant to activate codes, 201, 301, 401, 217, 666, and 888. His assistant complies, sending out the codes.

The ladies are sitting inside the CIA building when the alarms on their watches are activated. They are all in the undisclosed location of the CIA building.

Director O'Hara gets ready to explain the truth. As he walks around the huge, long white table, he puts his arms behind his back interlocking his hands and putting his head down. While walking in a slow pace, as to control his voice, he attempts to slowly unravel the story, in his mind, that he was about to unload on the ladies. Hoping he could make them understand the reason for the actions that they were taking to ensure the safety of the project at hand, so many years ago. Director O'Hara looks up at the group with a blank look on his face. Searching for the words to use to explain the CIA's latest dilemma—the Shadow Lady.

Director O'Hara reluctantly starts the briefing, and the girls looked baffled but eager to hear what Director O'Hara had to say. He walks around the table, then looks at the girls and then looks at Chief Williams.

Chief Williams looks away from the director and looks at the girls. He has sympathy in his eyes. They all see it and are even more eager and worried at the same time.

Just before Director O'Hara begins to speak, Silva interrupts him saying, "I'm sorry, sir, I been trying to hold it but, before you start, I have to go to the bathroom." Elise and Foxy started to giggle. Director O'Hara allows Silva to go do her business; she returns moments later. Director O'Hara speaks out loudly to everyone, "Does anyone else have to use the bathroom?" Then five seconds go past, three other people get up quickly to use the bathroom. Then Director O'Hara says, "Really guys, please hurry up, we have a lot of material to cover." Everyone returns to the conference room and took their seats. Then Director O'Hara finally begins to speak in a soft tone saying to everyone in the room, particularly the girls, "I want to start off by telling you ladies a story. Back in 1926, not long after the First World War and fifteen years before the Second World War, at least by our accounts until we were involved, there was a meteor shower. To be exact, the date was August 01, 1926. NASA tracked what would become one of the rarest meteor find in all of history. What made it so spectacular was the power it yielded. It was the size of a Volkswagen Beetle. They deemed it non-threatening due to the radiation levels were very low, but if they only knew the impact it would have in today's society, they would have blown it to shit when it first arrived. My grandfather John Lee O'Hara, a rear admiral in the US Navy, witnessed these events, told in his writings he recorded in a personal journal. These events based on my grandfather's accounts were deemed no threat at the time. It all started on a very cold winter night in January around two a.m. in the morning; the meteor, code name White Horse, entered our atmosphere, just off the east coast in the Atlantic Ocean. It traveled clear across the country. Traveling at 3,750 mph. Until it hit the Pacific coast, just outside of the Hawaiian Islands. In less than one hour, it landed about 75 miles off the coast of Honolulu.

"The US Coast Guard were the first responders arriving at 2:40 a.m. on the scene. They noticed an orange glare coming from the water. The captain of US Coast Guards, Patrick Johnson, became suspicious of the object, thinking it was a weapon from a foreign country. He immediately pulled his men back away from the glowing ore and ordered his XO to notify the US Navy.

"Upon receiving this crucial information from the US Coast Guard, my grandfather, Rear Admiral O'Hara, arrived on the scene around 3:05

a.m., just twenty-five minutes later, and relieved the coast guard. The admiral briefly detained the members of the US Coast Guard.

"Carefully he debriefed the men by advising them what they saw that night did not happen. The men understood exactly what the admiral was asking; like good soldiers, they agreed.

"As the mission proceeded, the US Navy started to receive and follow protocol procedures. They engaged the foreign object by using steel cables in order to extract the meteor from the sea.

"Special command informed the soldiers participating in the special operation named *Eagle Eye*. The crew successfully attached the steel cable to the meteor; without knowing the threat level of the object, the US Navy automatically deemed it hostile. It would remain hostile until proven otherwise; otherwise never came in my opinion.

"They witnessed a strange occurrence that baffled their understanding and defied all physics according to mankind's understanding.

"They noticed the steel cable began to change into a very shiny silvery substance that was not from our planet. The metal was unknown. Without knowing exactly what the *new substance was*, we pulled back from the object. Thinking the source had infected our metal with some sort of alien bacteria.

"This silvery substance had unknown properties that could not be explained by our top scientists.

"Little did we know this discovery would revolutionize mankind's existence to a whole new beginning, from which the world has never known!

"We named it Titanium.

"By using the sample from this new titanium metal, we infused iron ore. Which is used to make steel. Mixed it with the titanium. Then we implemented high frequency sound waves of sonic burst, in short increments using p-waves in a vibrational and longitudinal state. I know you do not know what I am talking about, so in layman terms, we created for the first time in history pure biological energy in a liquid.

"All this came from using samples from the meteor. That is when we knew this material was not of this world. Upon further testing of this material, we found out that it had life-changing properties. I will get into that a little bit later, so anyway our discovery was something unique and mind-blowing. Paved the way to creating a lot of different kinds of medicines that has changed a lot of lives, not to mention the different array of rare metals that came from our research using the meteorite. For example, diseases, human disabilities, deformities, old age, and emotional distress and a host of other things I have not mentioned could have been completely eradicated, but due to mass hysteria, the government thought that would happen. The higher-ups held back on the discoveries.

"For some mystery, the properties inside of the meteor gave mankind godlike power sort of speak. I will explain that a little bit later also. Our government and our allies partnered up together to form the Columbia Group. We started to set up laboratories all over the world, running test after test. It got so bad everyone trying to be the first one to have a major breakthrough. They became complacent many times miscalculating the calibrations, which proved to be fatal at times, causing catastrophic casualties. Like the blackouts in 1965 New York City, 1977 Northern Canada, 1978 Thailand, 1989 Canada, 2003 Italy, 2005 Indonesia, 2008 China, 2009 Brazil, and 2012 India. These countries all were experimenting with the white energy. The Columbia Group made up cover stories to hide the truth. We've also been experimenting on human subjects since the early 1950s. The 1954 experiment, in particular code name the White Horse, was spearheaded by one of the greatest minds of our time; his name was Professor Lee. His research was so top secret, presidents didn't even know of his existence. He would go on to invent a liquid solution; by using an ingenious algorithm, he converted meteorite energy into a liquid white energy. That would soon become a nightmare from then on. That would go on to affect me and everyone in this room and so on. A situation we were not prepared for.

"They noticed changes in test subjects less than an hour after exposure. Early on, it caused the male subjects to devolve back to primal and caveman-like instincts. More so than the women. 95 percent of all the males after being exposed died twenty-four hours later. Without any warning, they would experience cardiac arrest, then a massive heart attack would occur.

"Our greatest scientists could not explain why. We all knew deep down inside this thing we were tampering with was obviously of extraterrestrial origins. It could not be deciphered to man's true understanding at that time. Still mankind will never waiver, trying diligently to find the secrets to the elusive mysteries that haunt us without mercy. Trying to find the answer that has plagued us for years.

"Our scientists came up with the second phase of our experiment called *Project White Horse II*. Which ended in a complete failure. The experiment was scrapped. We lost a lot of people that day.

"This topic will be explained to you ladies at a later time.

"Now we need to focus on the blister that has come to a head. We need you ladies to burst it into oblivion.

"By taking down the very person who is trying to restart this nightmare. We call her *the Shadow Lady*, a.k.a. *Momma King*.

"Before you speak, I see your hand, Foxy. Please wait one moment. I will address all of your questions in due time."

Director O'Hara looks over at Elise, "May I please continue?" he says.

Elise looks at Foxy and Silva. She could see the confusion in their eyes. But being the leader of the trio, she knew she had to hold it down. Even though she was astounded about what she was hearing! She contained her emotions and gestured with a sigh of confidence to the girls, as if everything will be ok.

They both look back at Elise. In compliance, their facial expressions said to Elise, *we trust you*, and they sat back in their chairs. But still held on to their contempt toward Director O'Hara.

Foxy and Silva both, simultaneously, looked over at the chief with mean smirks on their faces of disappointment in their eyes.

Still trying to control and reassure themselves everything was fine. But it made little to no sense to them. So they continued to listen hoping to find out how they play a part into this whole, *Dr. Whos 'Space Odyssey,* story the director was telling.

So, Director O'Hara continued to explain the details regarding the *White Horse I & II Project.* Everyone's attention remained focused on him as he continues by saying, "I know this is a hard pill to swallow. Please bear with me, I am getting to my point. Professor Lee came up with many ways for the formula to be administered, liquid injections and even pill form. The only way to know of its presence it gives off a citrus smell from your mouth, after entering your bloodstream. But it could not be replicated through synthesizing the blood from a tested subject, which tested positive for white energy. For some reason, it could only be transferred from one person to another by birth or injections only.

"Our government wanted to know of its mysteries. We selected the best of the best candidates. We started with thirty volunteers. Fifteen males and fifteen females."

Silva jumps up as if she knows where the conversation is going. She rudely interrupts the director by yelling, "ARE YOU SURE ABOUT THAT? WERE THEY REALLY VOLUNTEERS, SIR?"

The director looks at Silva, then glances at Elise and Foxy, noticing their facial expressions. He knew right away they too agreed with Silva's question. They were all waiting for the director to reply to Silva's question.

Silva stood up, then quickly sat back down, trying hard to be respectful, but still showing her shitty attitude very boastfully in a sarcastic way showing very clearly she was displeased on what she is hearing. Director O'Hara attempts to answer the question but before he does. Director O'Hara glances over at Chief Williams. Before the director could reply, the chief quickly blurted out, "Keep a lid on it until the director is finished!"

They all complied, but not without reacting like high-school kids. They began throwing their hands up and rolling their eyes. Then they folded their arms across their chests.

The director answers all of their questions, "Yes, ladies. Everything was done by the book. We performed physicals. We asked health questions. We even had psych evaluations and tolerance testing. I mean everything according to the data reports I have poured through, including my grandfather's journal.

"After the tests were completed and the results were submitted, we selected our final six couples. Six males and six females. "They were taken down to a more secure area. Their environment safely controlled." The director was interrupted again, this time by Elise. Elise says, "Where is this story going, sir?" The director replies, "I'm getting there! Anyway, like I was saying." Director O'Hara continues, "We moved the final twelve to a more secure area, guaranteeing zero contamination. The professor met all the candidates, got them situated in their rooms, and started the injections. There were three phases, lowest dose, then medium, and finally the highest dose given every 48-hour increments.

"In between the injections, they underwent physical training and mental aptitude testing. The men showed slight elevation in their activities, which was a fantastic discovery. Except when you compared the females' test trails to the male test subjects, the women scored way above their counterparts in record numbers.

"Among the tests included hand and eye coordination. They were performed in tiers. As the tests progressed, these candidates experienced phenomenal powers that mainly showed up in high stressful situations. Then toward the end of the experiments …"

Foxy interrupted the director asking, "How long did this experiment last?"

"Anywhere from three to six months. When the male candidates started experiencing primal instincts, they were becoming very, very aggressive, at times. Then they would snap back as if nothing had happened. Then some of the men started dying one by one of cardiac arrests like I said before. We tried frantically to figure out what was happening, an answer we would never find. The strange thing is the women remained unaffected. In fact, the ladies were showing remarkable results totally opposite the men up until the end of the trail that ended tragically."

Chief Williams could see the anger within the girls and spoke out loud saying, before the director could continue, "Ladies, I need you to understand everything this organization did and does. It is for the security of this great nation we call the United States of America. We need you to know and understand that if nothing else. Ok?"

They look around.

"I said ok! You guys hear me? I said Ok?" says the chief loudly.

The ladies look at the chief and reluctantly agree with lazy tones and whispers for the third time, "Ok." They all replied with no confidence at all. Director O'Hara continues, "When we found out the side effects of the experiment, Professor Lee ran tissue samples on each of the candidates. He discovered that the male candidates' muscle tissues had been deteriorating in a fast and profound way that could not be explained. The white energy, which had been administered on three separate occasions. Something remarkable happens. The white energy became the primary nourishment of the host upon its introduction to the body. Replicating female cells so dramatically, on paper they looked like gods.

"When we found out that we were in the deep side of the pool. A place that we knew was out of our realm of understanding. Light years ahead of our time. We had not a clue of how to explain it!

"Professor Lee thought that this was an abomination, and he refused to continue or allow any more test subjects to go any further with experiment.

"I was a wet behind the ears, rookie recruit, fresh out of the academy at that time. This was my first top secret assignment. I witnessed one of the candidates who had found out about the cancelation of the program. She had become very enraged, for some reason.

"She was part of the original twelve. She was belligerent and physically outraged! She began striking members of the professor's team. Before they could subdue her, but only for a moment. She killed two females supposedly that tried to stop her from her rampage; they were candidates of the program, and Professor Lee tried to talk her down but it did not work. She had allies on her side that helped her; between the guards and the candidates fighting for control, it was complete chaos. The only good thing that came out of the catastrophe was the birth of six beautiful little girls.

"Three of the baby girls we lost. They simply just vanished; no one knows what happened to them according to our records, but the other three were rescued."

Silva looked surprised and confused. She says, "What do you mean three little girls were lost and three were saved? What does that mean, how do you not know what happened? You're the fucking CIA! I'm sorry, sir, but this is crazy."

Section Chief replies, "I know how it sounds. A lot of Professor Lee's work was destroyed in the explosion." The director replies, "The six candidate couples were always close to each other. They were volunteer couples that needed extra money because they wanted to start a family and the program exploited that. Those were the ones we catered to. They were also true patriots to the flag; that was a bonus. After *Project White Horse II* was dismantled, three beautiful baby girls were born out of the ashes that we had to worry about. So some colleagues and I decided that it would be best to help give them a possible life of happiness somewhat sort of speak. They were raised by trusted assets of the CIA. There they stayed hidden in plain sight until they were of a young adult age. Elise says, "What happened to the girls, sir, do we know where they are now?" Director replies, "Yes, we know exactly where they're at. I will explain that a little bit later as well; just bear with me, girls, please listen. I am nearly to the end. There was a woman who stood out amongst the rest of the test subject survivors. She escaped but not before killing the professor and associates that knew about the project."

The ladies raised their hands, and the director knew what they were going to ask at that moment. He quickly interrupted them and said, "Those names are classified until further notice. That order was given by someone even above my pay grade."

Director O'Hara continues, "Anyway, she escaped from the location. We had no idea how she was able to escape at that time. But by the time we found out, it was too late! We checked the footage. She blacked out right in front of the cameras; no one has been able to figure out how she did it. That is how she got her nickname the Shadow Lady. There was absolutely no footage to go on.

"We thought we could send a navy seal team after her. She was much more driven than they were. The result, she alluded all of us. As she took out more than half of our elite team of seals. We knew then what we created went far beyond our wildest dreams. The perfect killing machine relentless in every way. We searched and searched until we found out that she was being aided by individuals in high places, nothing concrete but our superiors they knew. We just could not prove it she has a lot of people on her payroll. Well, not until recently, we sent in an operative to go undercover inside the Houston Police Department. We had word that she was recruiting policemen, lawyers, and judges behind closed doors in the Houston Police Department. We have been set up here in Texas ever since. We went fishing and got a bite and hooked Momma King once again; been almost fifteen years since our last encounter. We've also found out she's been a very busy bee trying to monopolize the drug trade, the sex trade, and even the real estate markets, gobbling up every prime piece of property near all popular cities, all across the world. Do not really know exactly why but we have some thoughts. The Columbia Group became threatened by her presence all over the world, so they created The Network a joint task force that consisted of five countries: the USA, UK, USSR, CHINA, and JAPAN. The Network shortly after its creation eight months later was disbanded, due to agents being compromised because of bribery, then no one could trust each other. We were back at square one. The agencies from around the world became so paranoid, we all agreed this energy was too powerful for one nation to own. That is when we separated Professor Lee's algorithm into five distinct pieces that could unlock the secrets to the meteorite, thus creating white energy. We found a master craftsman, affiliated with the Columbia Group. He designed five beautiful music boxes: each one totally different than the next. They were entrusted to five of the most trustworthy capable assets each nation had to offer. Individuals we knew would protect these most sensitive secrets with their lives.

"Until one day ago, we found out a Chinese operative went missing and no sign of their music box was found." Almost on cue, a guard shows up to the door. The director waves him in. He whispers in the director's ears. Then leaves.

Director O'Hara turns back toward the ladies. Elise, Foxy, and Silva almost as usual, simultaneously all ask, "What was that all about?"

O'Hara replies nervously to the ladies, "I have a little bad news. As if I didn't have enough to give you already."

Elise says, "I don't think it can be any worse. My mother is dead. My sister went missing. What could be worse than that, Director O'Hara?"

Then Director O'Hara looks back at Elise with such concern in his eyes. Then he says to Elise, "There is a lot you all don't know; I'm just scratching the surface. The Chinese asset that was missing, we have just found out the identity. It is a woman that we are talking about that had one of the music boxes; she was found dead moments ago. Her name was Mrs. Lee."

The ladies wanted to know, "Who is Mrs. Lee?

The director replies, "You don't know her by the name of Mrs. Lee. You know her by the name Mrs. Chan."

The ladies' reaction was puzzling. They were frozen in disbelief that Mrs. Chan could be and was a secret agent. They became quiet for a minute. Trying to wrap their brains around the director's last statement. Briefly in shock again, seemed like the tenth time to the girls. Even though the girls still muster up the will power to hold their obvious contempt for Director O'Hara and Section Chief Williams, finally they come back to their senses. Then once again, one of the girls break the silence. It was Silva.

Silva angrily says, "What the hell! I am just going to say what we are all thinking. Why now! I mean why wait so long to tell us about this, obviously a maniac. That has been on our radar for decades, and we are just now hearing about it now, what changed, sir, and now you're telling us that Mrs. Chan is really Mrs. Lee from the donut shop? We've been going to her for years for breakfast!"

Director O'Hara replies back to the ladies, "I'm not trying to say what your hearing is easy but necessary; due to recent events, the threat is not going away. I am telling you what was given to me on a need-to-know basis. This comes from higher-ups as I have said before, so I was doing what I was instructed to do. But for what it is worth, I am so sorry you had to find out this way." Elise says, "So if Mrs. Chan was Mrs. Lee, so that means she was Professor Lee's wife right." Reluctantly Director O'Hara replies, "I'm afraid so, girls."

Foxy says out loud, "This is some fucked-up shit, Boss! What the hell is going on around here? How could you people keep this kind of shit from us! We are your best operatives. At least the last time I checked!"

Director O'Hara and Chief Williams quickly huddled whispering to each other. The director disagreed with the chief on something.

The chief replies back to the director in haste, "Sir, I respect you very much. But these ladies, I have grown to love like my own daughters, they deserve to know everything now since the chicken is out of the bag. So please continue, sir. Or I will. You can fire me now but they will know the truth about everything. Right now. Today! Sir."

The director looked at the chief and says, "OK. Maybe you are right. Ok. Ladies. Could you please sit down and refrain from interrupting me anymore? If you want to know the whole goddamn truth! It is not pretty. I tell you that!"

Foxy says, "There is nothing pretty in this line of work, sir. So please let us keep this show on the road."

Director O'Hara continues his story about everything, and I mean he explained everything to the ladies.

Foxy just kept saying to herself, "The donut lady. Seriously? The goddamn donut lady? This shit is crazy. This is so hard to believe."

Silva sitting down quiet. Still tripping from what her ears were hearing.

Elise gently puts her right hand on top of Silva's shoulder as to comfort her. Desperately the girls try hard to stay focused. For the second time, Foxy went on and on with a pissed off rant saying, "The donut lady? For real? The donut lady, what the hell?"

Then Elise said to the director, "So why was she in a donut shop?" Director O'Hara replies, "Well, her family owned a donut shop back in California, so pretty much the situation wrote itself. We thought that it would be safer for her to take back her family maiden name to protect her from any outside threats; at the time, it seemed feasible. Unfortunately, as you all can see it had a shelf life. We thought her having a donut shop wouldn't raise any eyebrows."

Silva came out of her shock, as if she hadn't heard a word the director just said, "Well, that didn't work out too good now, did it? And why was she in our neighborhood, Director?"

The chief interrupts briefly, "That's why we are here today. To clear up a long awaited wrong that should have been explained to you ladies a long time ago. And on my behalf, I apologize deeply. But I was under orders just as the director was as well. I hope you can understand, and I hope from the bottom of my heart that you can forgive us. For what you are about to hear next." The chief looks at the director and gestures for him to explain the rest of the story.

The director started off by apologizing in advance, trying to assure them everything that was done was all done for their protection. And the protection of the world.

Elise looks up at the chief as if she knew but she really did not. She put her arms around the girls once more to console them, in anticipation bracing for what would come next. In that moment, they were still upset and confused about everything entirely.

Then the director started by saying again, "Ladies, I'm really sorry. I was under orders like I have said many times before. I hope that you can understand that."

Elise replies in a roar of a voice, "JUST TELL US!," then instantly she calmed herself down, then whispered to the director, "OK, sir. I understand. Protocol. Start talking. You're killing me with the suspense!"

The director replies, "Ok. Well, ladies, as I have said before, there were five people we entrusted with the music boxes. As you know Mrs. Lee was one, but what you did not know is your mother was another one. Well, the mother you know as your mother."

Everyone stood. In shock again! Foxy replies, "The hits just keep on coming. I know I am going to hate myself for asking this question, but here goes nothing. What do you mean by that Director O'Hara?"

"Well, you wanted to hear the truth! Let me finish, it gets worst!" says the director.

Silva says, "Damn it! I feel like we need helmets for this hard-hitting shit you guys throwing at us!"

Foxy asks in pain, "Are you serious, Director?"

Elise chimes in, "Man! FUCK! This is a fucking nightmare!" Then she calmly looks back at the chief in disgust. Then she looks back at the director and says to him, "Keep talking; don't leave nothing out."

The director continues talking with regret in his voice, "Elise. Who you thought was your mother is not, she was your aunt. Your sister is really your cousin?"

Elise's eyes turned bloodshot red! She jumped up yelling, "YOU'RE LYING! THAT'S A FUCKIN' LIE! HOW IS THAT EVEN POSSIBLE? SEE Director, YOU'RE GOING TOO FAR NOW!"

Silva and Foxy, trying hard to stay planted in their chairs, their first thought was to destroy everything inside that room and anything else that got in their way.

The director continues, "Elise, Silva, and Foxy, your mothers and fathers were six of the twelve candidates that volunteered for the program."

Foxy jumps in quickly saying, "So you're telling us that we were abandoned by them?" Director O'Hara replies, "NO. That is not what I am saying. Please let me finish. Foxy and Silva your parents were killed trying to stop the Shadow Lady, a.k.a. Momma King!" replied the director.

Silva says, "Who in the HELL is Momma King? And what does she have to do with our parents' death?"

The director looks at the ladies with an expression of sorrow, which was written all over his face, "I know it's a hard pill to swallow right now. You have to accept it. We can move on, ladies; you have to be strong. You were trained for misdirection."

Suddenly, they stop crying and jump up out of their seats lunged toward each other to comfort one another in a loving embrace.

Even though they all were a wreck, and for good reasons, they slowly started to register that their parents were murdered, and they were lied to by the very organization that they vowed to protect with their lives. For love and loyalty of country.

As Elise stood to her feet, she looked around the room thinking to herself, in graphic details, about putting a bullet in the director's and Chief Williams's head!

Then she snaps back, looking at the ladies, shaking her head thinking to herself the name Momma King understanding that she was the one responsible for her life turning to shit.

She regains her composure, then says to the director and Chief Williams, "So, now tell us, RIGHT FUCKIN' NOW, Who IN THE HELL IS MOMMA KING EXACTLY?"

The chief looks over to the director as he has been doing all day since the meeting started.

Foxy speaks up before they can say anything and says, "Well, Boss, you've said a lot. Now finish it, so we can get to work, sir."

The chief gets up out of his chair as Director O'Hara sits. Section Chief Williams begins by apologizing, again, for not disclosing this information sooner.

The ladies understand and are waiting impatiently for the chief to continue telling them what is going on.

Chief Williams says, "The woman that is responsible for everyone's demise, her government name is Cherrell Katera Bishop, a.k.a. Momma King." He looks over at Elise. "Yes, Elise. She is your mother."

The ladies react in horror and shock. Not saying a single word.

Elise gets up and walks to the entrance of the door.

Silva tries to follow, Foxy pulls Silva back, then whispers to Silva saying, "No, she needs some space in order to process this. Give her a moment."

By now Elise is screaming and stomping the floor very heavily, kicking the walls just outside the room in the hallway! The guards heard the noise and commotion and hurried to the location of the noise to render support or something. As the guards showed up, Director O'Hara motions them away.

After fifteen minutes, Elise calmly comes back into the room. Sits down not saying a word. No one spoke to her directly. Then she looks back toward the chief and asks, "Please continue."

He does, "We've identified the contents from the package you ladies brought in. The three items are Elise's sister's iPhone and a note written to you Elise and your mother. We do not know at this point if it was written by her on her own free will. Or was she manipulated into writing the note."

Elise quickly replies, "What did the note say, better yet, let me see it!"

The chief called in the item so the ladies could examine the contents.

Elise says, "This is my sister's handwriting and her iPhone. But the message does not make any sense to me. Do you think they have my sister, sir, for a ransom?"

Then Director O'Hara replies, "Well, if we have received no demands yet, we've been running tests on the iPhone and the note so far no leads. The note says, *Do not be worried. I will contact you all later. I love you all. Love, Kawanna.*

Elise says to the director, "What's the plan?"

"I thought you would never ask!" replies the director who had been anticipating that exact question.

"We had an asset inside the Shadow Lady's organization. For a while now," the director continued.

"What is a while, sir?" asks Foxy.

"About eleven months, as of twelve to fifteen hours ago. His cover was blown, beaten nearly to death, and left for dead on the cold city streets of Houston to bleed to death, but not before he could get us some vital information about the Shadow Lady's realm of operation. Thinking he was going to die, he sent us a letter through the mail. With the help of a double agent named Rachel Agent 888 was able to escape," replies the director.

Silva asks, "Well then, where can we find him?"

"I got one better than that," says the chief. He gets on the phone telling the person on the other line, "Send them in now." Then he hangs up the phone.

Two people walk into the conference room. One was Detective Skaggs and the second was Rachel Chong.

Elise speaks to Detective Skaggs asking, "Do I know you? You look familiar."

Foxy and Silva both agree with Elise.

"You should. I am Detective Skaggs," says Detective Skaggs. I was the one at the morgue, remember.

The ladies comment at the same time, "Yea, that's right. You are Detective Skaggs. Do you know what Momma King looks like exactly?" Agent 888 replies, "Yes, I do. Let me explain the whole layout of the organization and why."

"I was placed on the police force because the agency were tipped off that Momma King likes to recruit corrupt cops to work for her. The agency saw this as an opportunity to finally figure out the Shadow Lady's true plans, so they gave me, Agent 888, a nasty bad boy profile through and through, creating a bait for her, then she bites after four-and-a-half weeks. I was approached by Darius Khan, one of Momma King's deadly henchmen. Momma Kings #1 highest assassin in the Red Scorpion Clan.

"He was set to protect her from an unknown source that is a whole other story. We have yet to identify his boss. Every photo we have ever taken was taken out of focus. But we do know he is of African descent. That is all we know."

Director O'Hara says, "Well, I have my best team working on this case, with a new addition to the team, Rachel Chong, Agent 888, and my ladies. I'm hoping we can find all the answers to all our questions to bring this sadistic bitch down once and for all. I yield the floor back to Agent 888, a.k.a. Det. Skaggs. He will brief you all on the latest occurrences to prepare everyone for the mission. Stay sharp, guys, we go hot at zero six hundred." Detective Skaggs briefs the ladies once again on everything he knows about Momma King and her operations.

CHAPTER 20

The Waiter

Six forty-five in the evening, there is a waiter pushing a cart down a hall-way. Coming out of a kitchen, he passes a slew of naked women walking past him. They started grabbing on him and making strange noises. They were flirting with the uncomfortable waiter. He just kept his head down and continued pushing the cart nervously to its destination.

Walking quickly, he arrives at the end of the hallway. It was kind of dark; there to his right and left were two sets of elevators. The waiter stood in front of the one on the right for thirty seconds. After he pushed a white button, the red elevator door made a ping sound. Then the doors opened. The waiter gets on the red elevator and he quickly pulls the cart onto the elevator. The door pings again and the elevator door closes. Instead of the elevator going up, it went down. It seemed like fifty floors, but it was only twenty floors down. The elevator stops and the waiter gets off the elevator and the door closes.

He walks down a dark corridor that was lit by red lighting. There were double doors made of red marble with lionesses carved on each door. The waiter pushed the cart through the doors. There were three people sitting around a very large and a very impressive handcrafted red wooden table.

The table is so beautiful, it renders you speechless when you see it for the first time. What got the waiter every time was who he came into the presence of sitting behind the table. He was always in awe, wondering to himself who could construct such a beautiful table. It was hand carved with an African Egyptian queen placed in the middle of the table with Amazonian warrior-like women carved throughout the rest of the table with a very shiny glaze, almost looked like it was wet.

There were two men and one woman. The first man was seated closer to the door at the end of the table opposite the woman. It was Darius Khan, of course the Shadow Lady, a.k.a. Momma King. She was sitting the furthest away.

The other man was sitting in the middle of the table. He was dressed in a snow-white silk suit. With fur around the collar and around his wrists, two platinum bracelets. The bracelets had a wolf head on each of them. The man dressed in white stood up for a moment to straighten out his pants. He had on a pair of polar bear skin shoes laced in platinum, if you can imagine that, and a white hat with a rare white peacock feathers, seemed like they were perfectly placed in layers with skilled precision on the side, and when the sunlight hit it at the right angle. The feathers give off a beautiful array of colors, like a marbled rainbow. He had on a pair of white and silver sunglasses. Whether he was inside or outside, he always wore those sunglasses. The only time he took those shades off was to clean them.

He stands about six feet, four inches. His skin is like golden red fire mixed with burnt bronze with a flawless smooth complexion. With a Billy D. Williams vibe and a Denzel Washington swag, long permed hair, at least it looked like it was permed. He spoke with a high pitch, a mix between Chris Tucker and Mike Epps, yeah imagine that. He also spoke with a (ebonics: "Jive talk" black American slang used mostly in the seventies) voice something like a pimp. He walked with a long platinum cane. The bottom and top of the cane was made of white gold and platinum. The top of the cane was in the shape of a wolf head that was attached to a hidden custom Japanese sword. The wolf head also matched his white diamond platinum bracelets, but to the normal person, they appeared to be jewelry. When he flicked his wrist up in the air, the bracelets would fly off his wrist up in the air turning into very sharp, very lethal daggers, then fall back into his hand in the blink of eye. His name is Shawn Brown, *the Pimp*, another one of Momma King's top cohorts.

The waiter soon finished passing out their meals and pouring the wine into their glasses. Then Darius Khan motioned the waiter to leave. As quickly as he showed up, he was gone just as fast.

Shawn Brown, the Pimp, looked over toward Momma King to inform her that the secretary of defense was on board with their plans.

Momma King shows a delightful smile. She says, "Make sure everything goes according to the plan. I don't need any more failures."

The Pimp agreed. Then they finished eating.

Darius informs Momma King saying, "My master will be attending the meeting set for later this week after the package is taken."

Momma King replies, "Wonderful. Finally, some good news. I can drink to that. Now, we have to do is find the last three music boxes. Thanks to the *girls*, for leading us to Mrs. Lee. Without them even knowing!" Thinking to herself, Why haven't the CIA told them about her yet? She celebrates to herself that Mrs. Lee is finally out of the way. Then Momma King says to herself in a thought, If my sister would have taken my side, I would have spared her life, but she was too weak. Letting her emotions get in the way of a greater purpose for all women. She had to die for the cause, like that her thoughts were over. Then Momma King says, "Darius, I need to go ahead with the plans we've set in place. Shawn, I need you to keep your minions on the CIA, especially those girls, and I repeat, don't kill them. Just apprehend and bring them to me, is that clear? Everyone else I could give, a shit, make it clean and quick and only if they deem a threat."

Shawn replies, "Yes, mistress. I understand." Momma King says, "I want to know their every move and their plans. I know they're concocting something, like mother, like daughter. Find out what it is. Now leave."

The Pimp and Darius leave Momma King's presence.

CHAPTER 21

The Mission

Meanwhile, back at the CIA Headquarters, Detective Skaggs, a.k.a. Agent 888, finishes his briefing. He takes his seat.

Director O'Hara stands up, a guard comes in, and hands him a stack of papers. Then he exits the room.

The director looks over the paperwork, then he gasps! He looks up and around the room at everyone and says, "Well, everyone, we have our orders. The first order of this operation, according to Detective Skaggs's briefing, and what our top minds upstairs have come up with, are as follows: we need to infiltrate Momma King's lair. It is located, as far as we know, right under a gentlemen's club according to our intel from a waiter, code name Blue Egg; that Agent 888 Skaggs had as an informant. He was pulled out and placed in witness protection. As of thirty minutes ago, Blue Egg and four US Marshals were found dead at a compromised safe house.

Foxy chimes in, "What? This bitch is serious!!" Elise replies, "She better be. Let it be known, fear is a stranger here. Never knew the bitch and will strangle a bitch."

Foxy, Silva, and Elise look over at Rachel.

Rachel says, "I think we're going to get along perfectly. We will all get revenge on that bitch, Momma King, for destroying our lives or die trying!" The girls, Detective Skaggs, Chief Williams, and the director all have a special saying they learned at the academy: they chant eight alphabets, which is eight hidden words that form an acronym, once in unison with aggression, A.G.T.A. A.T.T.A. (AMERICA GOT THAT ASS, AMEN TO THE AMERICANS). You pronounce it, Aa-Gee-TA-AT-TA. It has a two part meaning: (1) to agitate (Italian origin) (2) chiefly US (Hindi origin).

Silva blurted out before they could finish, saying in a southern belle country-like accent, "Is this bitch for real (ebonics: y'all)? I'm so serious, my body hurts. I need an aspirin, shit. Goddamn muthafucka, (ebonics: imma) fuck (ebonics: dat) hoe up! Man! I cannot wait! With her (ebonics: stanken) ass!"

Then she caught herself, damn near forgot where she was, then quickly shuts up!

But not before she says, "All shit. Sorry about that guys. I had an insane in the membrane moment."

The director looks back at her perplexed. Then he quickly changed his expression into an apologetic tender kind expression, like okay whatever that means. Then he says, "You are right, Silva. Trust me I feel exactly the same way, I think. This mission is mainly to set things right. This is a vindication for you guys. I agree 100 and 10 percent on fucking that bitch up too. Everyone was totally surprised to hear the director talk like that, then all started to clap their hands in agreement. After the director's remarks, instantly they were even more motivated.

Foxy quickly says, "Yes Boss, you go Boss."

The chief looks at Foxy in agreement. Then Elise says, "We all have a score to settle with pure evil that has plagued our lives. Let us just stay alive and get it done, ok!" Everybody agrees, then Director O'Hara finishes up the briefing by explaining how to take down Momma King. He introduces a few more members that entered the room at toward the end of the meeting. Hopefully, they will successfully complete the demise of Momma King once and for all.

Director O'Hara excuses himself from the room.

Section Chief Williams thanks the director for his time and proceeds to walk him to the secure elevator doors where they say their goodbyes.

Section Chief Williams walks back into the conference room. In that time, there were over fifteen heads inside the room. The extra heads in the meeting were highly trusted agents of the agency; their names were not disclosed for obvious reasons. The Shadow Lady had spies everywhere; these were the only ones they could really trust with the mission. They are ready and eager to start the mission. The chief started by saying, "I want to thank each and every one of you dedicated men and women here today. For representing the United States of America. I must inform you guys that this mission will be highly dangerous. Unlike any other mission you have ever faced in the past and I do mean ever. And some of you might not make it back alive!"

Then they all said in unison, "A.G.T.A. A.T.T.A.: Aa-GEE-Ta-At-Ta, sir. After that like the Marines, they all say, *WHO-RA!*"

Chief Williams replies, "I couldn't have put it better myself. Elise, you need to get in contact with you know who to get to be briefed. He is the only one outside the agency we can trust." Elise replies, "Yes sir, I understand my orders." Chief says, "Go with GOD and be safe. Now get out of here, you guys. All of you have your orders. Exactly at 1900 hours, Operation "birdcage" and "sleeping wolf" will commence simultaneously. May God have mercy on us all," then everyone leaves.

CHAPTER 22

Gonzo Ramos

Elise, Foxy, Silva, Agent 888, and Rachel report to their contact, at a roach motel named Rosie's. Where the CIA asset contact is. His name is Gonzo Ramos.

Gonzo Ramos is a small Mexican man. He is five feet, five inches tall. He has dark brown skin with big round eyes. At first glance, you would think he knew no English. But he speaks English better than they did, with no accent.

He is a familyoriented individual with a wife and four daughters, no sons. Many people think that is the reason what compels him to do what he does; he says if he can make this world one day safer for his little princesses, it is well worth his life. He believes in karma; his number one rule above all rules with one exception, dying for his family. Always remain very respectful of women. That is probably why the hookers gave him the nickname Mr. *Nicey*. Everyone in the neighborhood knew he was a cool cat. Over the years, he managed to fly low under the radar undetected for some reason or another. Always maintaining his anonymity.

That appealed to the FBI and the CIA. He is an informant that has been feeding information to the Feds and the CIA for over ten years. He could get inside places that normally others could not. That is what made him so valuable and the reason he has been on the government's payroll for so long.

Gonzo plugged them in with some street hookers that knew a few of the women who worked at the gentlemen's club called Sasha Erotica.

Sasha Erotica is owned by Shawn Brown, the Pimp. It is rumored that Shawn 'Pimp' Brown named the club after an old childhood dog. That is what some people say. He is known for jokingly saying to some of his closest friends that his female dog Sasha was the only bitch he could truly ever love and trust without a shadow of a doubt, hands down was the most loyal!

Back inside the room of the motel called *Rosie's Motel*. Right down the street in Third Ward, off NW Highway I-59. About three miles from Sasha Erotic, the gentlemen's club. It was located off 59th and Quitman.

The ladies settled into the room rehearsing in their minds. Then they started laughing, with dark thoughts of satisfaction, daydreaming of killing the Shadow Lady. While trying to act normal. The girls' hearts are full of vengeful tender. This was ready to be deposited into the Bank of Pimp Shawn Brown. Yielding forever payback, pimp-money residuals. Then the girls snap out of it. Everyone including Det. Skaggs and Rachel. All go to their post. At 1900 hours, everyone set the times on their watches. Then Operation "birdcage" and "sleeping wolf" commences.

CHAPTER 23

Phase 1

Friday night, 7:05 p.m., Elise posts up at her spot waiting on her *so-called* Tuesday night regular. Mr. Andre, an fifty-nine-year-old CIA-Operative, Vietnam Veteran. He loved him some Ms. Elise. *Elise's code name in the field is Lovie Nicole Cole but they just call her Nicole.*

Nicole notices Andre pulling up. He was driving a candy red CTS 2020 Cadillac. His window rolled down and a flock of women rushed up to his car.

Mr. Andre speaking with a slow greeting in his voice speaks to the women. Andre says, "Hello, ladies, thank you for your time but I already made my choice." They were trying so very, very hard to sell themselves.

But Mr. Andre pointed at Nicole. All of the other women dispersed from the car to go back to work looking for another John. Keeping up appearances they perform their roles perfectly as trick and the hoe.

Then Nicole says to Mr. Andre, "Hey, Big Daddy."

Nicole was wearing an ocean blue shiny mini skirt, with white stilettos with blue rhinestones embedded into the heel of the shoes. She had on a big afro wig, blue lipstick, long eyelashes, and two blue rhinestone bracelets, one on each wrist—there were also communication and location devices hidden inside—and a white-and-blue purse that hung to her right side.

The way she walked to the passenger side of the car, it was as if she is auditioning for a top runway model from the '70s. Andre thinks to himself, Oh boy, oh boy, this woman is so beautiful.

He blushes with a gentle smile on his face, then Andre says jokingly to Nicole, "Damn girl, you really look the part. You look like a fine ass hoe."

"Shut up, Andre." Nicole replies with a blushing giggle.

"But before we get started, I just wanted to say I'm sorry about your mother." Andre says with all sincerity.

Nicole replies, "Thank you so much. I really appreciate that, Andre. Now let us get to work. Do you have the asset in place?"

Andre replies, "Yes, I do. He is an old military friend from back in the day. His name is Mr. Remy." Then Mr. Andre pulls off from the curb. He started driving toward the freeway. He looks over at Nicole and asks her, "You comfy?"

Nicole says, "Yes. I'm fine."

Then Andre picks up his phone and makes a call to Mr. Remy. The phone rings once.

Mr. Remy picks up the phone. He knows who it is instantly. Mr. Remy says, "What's up, old man? How's life treating you today?"

Andre replies, "Ok. I am fine, Buddy-boy. I have someone with me who needs to talk to you for a moment."

Remy replies, "You know them well."

Andre replies, "Very well."

Remy agrees to a meet. Remy says, "Come over to my house. I hope you have not forgotten where I live at."

Andre replies, "What do you think?" Laughing under his breath, Mr. Remy says, "Well, see you in a few. Oh, yea, I am having a family over for a barbeque. So, don't let the people scare you away." Andre replies, "Ok," laughing out loud.

They both hang up the phone.

Andre heads over to Remy's house. He looks at Nicole saying, "He'll see us now."

"Well, ok, let's do it," Nicole replied.

Wearing a pleased look on her face. Thinking to herself, *Phase 1 is now in effect*. Then she picks up her phone and calls Foxy Roxy.

Foxy Roxy answers the phone on the second ring and says to Nicole, "What's up? What's the word?"

Nicole replies, "Go ahead with your shit."

Foxy Roxy says, in a playful way but still maintaining a level of seriousness, "Ok bitch, gotcha. On my way to the spot."

Then Nicole (Elise) hangs up the phone with Foxy Roxy.

Meanwhile as Foxy Roxy gets off the phone with Nicole, then calls Silva, a.k.a. the Tongue.

Silva the Tongue answers her phone saying, "What up, Fox?"

Foxy Roxy replies, "You already know."

Then Silva-Tongue says, "Ok gotcha. On my way."

Silva hangs up with Foxy. Silva quickly gets into character. She sees the alley in the back from where Nicole was posted up at. Silva sits and waits.

It was not long before *the mark* (Freddy) showed up. Freddie approaches Silva saying, "What is your name, sweet thing?"

Silva goes into character, flipping her hair back. Then she started sucking on her lips and rolling her eyes toward him. In a sexy hot way that women usually do to take total control of a man's undivided attention.

Then Silva-Tongue replies, "Who wants to know?"

The man walks up closer to Silva and says, "My name is Freddy Ready."

Silva-Tongue replies, "What kind of name is that?"

Freddy Ready replies, "Because I'm always ready for whatever!"

Then Silva-Tongue replies, by laughing and giggling. Trying to come off as dingy and unintelligent, Silva sells it, playing her part perfectly. Freddie looks at her as if she were easy prey. Little did he know that he would soon become her prey. Silva continues saying, "Oh, that's cute. I like that. On the fly, she thinks of a name to hook his curiosity and it works. Well, my friends call me Silky Tongue."

Then Freddy replies, "Now you got my mind wondering. Yes, I am going to ask why they call you Silky Tongue." Then he sees Silva licking her lips. Then she leans over toward Freddie Ready, then whispers in his ear saying, "Wouldn't you like to know? Let's just say if the day goes well, who knows maybe you could be a new friend to me." Freddie Ready replies, "Hell, I like that even more. By the way, it is usually more girls around here. Have you seen them?" Silva Tongue replies, "They're gone with Johns. I got here a little late; well, on the other hand, I guess I got here right on time 'cause here you are." Then Freddy Ready blushes, then he says, "Are you new? Are you one of Shawn's girls?" Silva replies, "Shawn who, that name sounds familiar but no I'm by my lonesome." Then Freddie Ready says, "Kool, that is what's up. I'll take you under my wing and show you the ropes, sounds good." Silva replies, "Sounds good to me, okay I'm in." Silva plays her role as the dingy, naïve but beautiful vixen with flawless precision. Then Freddie Ready says, "Silky, are you hungry?" Silva replies, "I could eat but I have a taste for some good old-fashioned bar-b-que. How about you?" Freddie says, "Yeah, I could go for some bar-b-que. Alright shit, let us go, girl." Silva replies, "I hear that daddy, so let us go. Where's your ride?"

"Right over there across the street. The cherry red Mustang, with the polo chrome rims, sweetie," says Freddy.

That is when Silva knew that he loved that car, a lot! Because the whole time he was talking to her, he could not take his eyes off that damn car. Not even long enough to see Foxy walk right behind them.

Freddy continues, "Well, let us go riding."

Silva-Tongue agrees. They proceed to walk across the street.

Freddy runs around the car to the passenger side and opens the door for Silva-Tongue. She gets in. Freddie thinking to himself he found a bottom bitch to claim before Shawn Brown got a chance to sink his teeth into her. Freddy closes the door, walks to the driver side of vehicle, and gets in. Take one more look at his prize sitting on the passenger side, so he thought. In a very confident way, he expressed victory that seemed to pour from his eyes. Then he starts the car and proceeds to drive off.

He speaks in a slightly loud voice, as if he forgot something, "Damn."

Silva-Tongue says, "Everything ok. You still taking me to the bar-b-que shack, right Freddy?"

Freddy replies, "WHAT? I didn't agree to that, did I?"

"You know what, that's ok, Freddy," replies Silva-Tongue. Silva never mentioned the bar-b-que shack neither did Freddie, but Silva had him so engaged in the conversation, she made Freddie think he did.

Freddy says, "That's ok. No, I'm a man of my word. I mean, I will take you. That is on the other side of town, right?"

"It's a little away but not that far," Silva-Tongue replies.

Freddy says, "Ok but I have to make a block right quick to take care of something. Then we will be on our way. Ok, Silky."

Feeling disgusted in her thoughts she replies, in a low whisper, but loud enough to be heard, "Cool, Freddy."

Freddy circles around the block. He gets out, then goes into the alley to check his stash house. Silva-Tongue stays inside the car. Then Silva talks on a secret intercom on her bracelet to Foxy Roxy saying, "Everything good, girl." Foxy responds into Silva's earpiece saying, "Affirmative." Then they quickly cut back to radio silence. Then a few minutes passed by, then Silva-Tongue hears an angry savage screaming sound coming from the back alley.

Thinking to herself, Yep he knows now. Staying in character as she asks Freddy Ready, "What's the matter, Freddy?" In a concerned way as she jumped out of the car.

Freddy was walking back and forth on the sidewalk with hands over his head, whispering little chants, "What the hell, what the hell!" over and over again. Trying to wrap his head around the shit he just witnessed. His stash house just got robbed and all his men at the house were all murdered.

Then Silva-Tongue offered up the bait. For the next phase of her plan, just to see if he would bite. Silva says to Freddy Ready, "Baby, when I came up, I saw a pimp-looking man walking out of the alley way with three of his guards. I think they were his guards; they had big ass guns and they got into a Cadillac Escalade 2021. I know that because I saw it on a commercial last night. Pearl white, shiny SUV. And he was chanting some stuff." Freddie replies, "What stuff was he saying?" Silva Tongue thinks to herself I got him. Then Silva says, "Stuff like fuck that nigga, I told his dumb ass?" Freddy looking confused, as he looked up at Silva-Tongue. Freddie replies, "So he said that. I know we've been having some territorial issues, but I never thought he would go this far."

Silva-Tongue says slowly to Freddy, "Oh yeah he also said, as he got into his vehicle, fuck that muthafucka, this my hood." Freddie in shock with disbelief written on his face. He just stood frozen in a trance. Then Silva-Tongue looks over her shoulder across the street. She noticed Rachel pulling up with Detective Skaggs in a blue-and-white cable van. Looking back at her as they prepared themselves for a long night.

Detective Skaggs gets up out of the passenger seat and walks to the rear of the van. The back of the van was filled with radars, GPS, and a slew of high-tech electronic equipment used for surveillance, phone interception, and video recordings. They could see out of the van but no one could see inside the van. Detective Skaggs checked the surveillance equipment to make sure everything was working properly. He got ready for his part in the next phase of the operation. Getting back to Freddie and Silva across the street.

Freddy Ready snaps out of his trance. He looks at Silva-Tongue and says, "You know what? That sounds like something that arrogant prick would say." Then he took out his cell phone and tried to call Shawn Brown, the Pimp. Foxy-Roxy is long gone from the stash house; the CIA operatives get into position.

Detective Skaggs intercepts the call and Rachel gets on the line. Pretending to be the Pimp's assistant and secretary, "Hello."

Freddy Ready replies, "Where the hell is your boss?"

Rachel replies in a deep, nasally voice saying, "Hello, Mr. Freddy. We have been expecting your call." She puts on an award-winning performance convincing Freddy Ready that everything Silva-Tongue said was true.

Freddy Ready was shocked as he thought to himself, Shawn Brown, the Pimp, really was the culprit that invaded his stash house. In that moment, Freddy-Ready wanted to know what the hell was really going on.

Then the lady came back on the phone and started saying, "Mr. Brown would like to meet you at," but before she could finish, the phone hangs up.

Freddie became enraged cussing and carrying on. Thinking he had a bad connection with the phone call. He tried to call back, but the phone kept saying there was no service in the area. He thought to call his crew, but before he did that, he did not want to act in haste until he knew all the facts.

Silva says, "Freddie, now I know why I wanted bar-b-que. I remember overhearing that pimp guy you call Shawn Brown talking about that new bar-b-que place that just opened up. He said to one of his bodyguards, word for word, let us go get some que from the shack, guys. All this work got me hungry."

Freddie replies, "He better have a damn good reason for this shitshow he created. If not, I will have something for his ass to eat on. Are you sure that's what he said?"

Silva replies, "That is exactly what he said. I have an excellent recall memory." Staying in character with a dingy high-pitched voice. Freddie Ready says, "For your sake, you better be right." Silva staying in character she replies, "Nigga, is that a threat? I don't need this shit, okay. I am trying to help you out. Because you seem like an alright dude."

She just kept going and going laying into Freddie.

Then finally Freddie says, "OKAY! OKAY! SHIT! I believe you, sorry! My muthafuken bad, alright!"

Silva acting like she did not hear him.

Freddie says for the second time, "I said alright damn girl, we good."

Silva still in character acting pissed off rolling her eyes and she had her arms folded up biting on her bottom lip laughing inside her reply, "Okay, I am good. Let us just go."

Freddie says, "That asshole does like eating some bar-b-que. That's got to be where the fucker is headed to, and so am I!"

Freddie still trying to call Shawn Brown, then Skaggs and Rachel answered with plates and silverware clacking up against each other creating a restaurant background sound. Then they hung up the phone and never answered it again.

That is, when Freddie became enraged and was hooked into believing the story Silva had told him even more. Freddie and Silva finally pull away from the curb heading straight for the bar-b-que shack.

CHAPTER 24

Phase 2

Fifteen to twenty minutes earlier, Foxy-Roxy dressed in all black comes up to a building. That is parallel to the building that is housing Freddy Ready's drugs. Foxy Roxy speaks into her watch com saying to Detective Skaggs and Rachel, "I am at the spot for the next phase."

Detective Skaggs and Rachel driving in a cable van, he replies, "Our ETA is ten minutes. Set your watch in 5 … 4 … 3 … 2 … 1."

Foxy-Roxy and Detective Skaggs say at the same time, "Set!"

Foxy-Roxy says, "I'm going in."

"Roger that," replied Detective Skaggs.

Foxy-Roxy pulls off her black backpack, pulls out a long black and chrome gun-like mechanism with a hook on the end and a scope with a laser marker on top. It has a digital screen that showed the distance from the ground to the top of the building in feet and inches. She focused, then aimed the gun-like instrument upward. Found her mark, then presses the trigger. The hook shoots out and up toward the roof of the building. Hitting her mark perfectly.

She scaled the wall to the roof of the building. Then proceeded to make her way to the next building from the roof. She located the air vent that was marked on the blueprints that she pulled from her backpack.

She entered the air vent working her way through the ventilation system.

By then Detective Skaggs and Rachel made it to their spot across the street from Silva-Tongue, just before Freddy Ready pulled up, which was five minutes later. Silva-Tongue kept him busy for the next fifteen minutes.

Detective Skaggs located Foxy-Roxy using an infra-red body heat seeking imaging laser to find her position, speaking into his watch com he says, "There are three men at the bottom of the stairs and two men guarding the second floor just at the top of the stairs. Four more inside a room on the second floor facing the stairs. Twelve feet exactly from the last step to the door to the room with the drugs and money. Inside the room one man is counting and the three are packing up the drugs." "Roger that," replies Foxy-Roxy. She pulls out her gas mask, then throws a canister of knock-out gas from the air vent. Completely immobilizing everyone who was in the room with the drugs and money. Then the two men on the second floor heard the noises coming from inside the room where the drugs are.

By then Foxy-Roxy was out of the ventilation system. The two men stormed into the room shouting; the first one yelling, "WHAT'S GOING …?"

But before he could finish his sentence, Foxy-Roxy, standing behind the door, grabbed him by his neck and quickly broke it! His body fell on top of the table filled with money and some of the packaged drugs. Then the table broke with him, and all the contents came crashing hard on the floor!

The commotion alerted the three men on the first floor … they are on their way up!

As the first man on the second floor fell to his demise, the second man grabbed Foxy-Roxy by the throat and pushed her hard slamming her savagely into the wall. Foxy-Roxy quickly kneed him in the balls, the immense pain caused him to loosen his grasp from around her throat! She pushed him off her and hit him in the face with the butt of her gun. That move dislocated his neck from his spine killing him instantly! All this happened within fifteen seconds of her coming out of the ventilation system.

By that time, the three men were coming up the stairs. Foxy-Roxy came out of the room and hid just on the side of the wall near the stairs just out of sight. As one of the three men made it to the top of the stairs,

Foxy-Roxy let him pass her. Then she pulled out her freshly loaded gun with the silencer, pulled the trigger. Putting two shots into the back of his head, killing him instantly. He slid face down across the floor into the wall.

The second man ran toward her, grabbed her by the shoulder, tried to swing her around. She immediately countered his move by putting her right hand backwards around his neck and flipped him over her shoulder onto the floor and placed two bullets into his chest; instantly he stops breathing.

The third man saw his friends lying dead in the stairwell in a matter of seconds. He looked in the face of death and it belonged to Foxy-Roxy. Then the last man not taking any chances turned around and made a beeline back down the stairs.

While on his way down the stairs, he was trying to get his phone out of his pocket to call for help. As he was running back to the base of the stairs, he was able to put the phone up to his head waiting on the phone to ring. Then there was a chopping sound. Then he began to fall to the floor. He had a chrome shiny knife stuck thru the back of his head, coming out of his mouth. While he was still holding on to his phone. He died before he hit the floor, leaning forward hitting the floor like a bag of potatoes.

Foxy-Roxy celebrated inside. Thinking to herself, *You still got it, bitch*!

Detective Skaggs says to Foxy-Roxy on her radio watch, "We have company!"

Foxy-Roxy quickly gathers the money and the drugs. She looks around, then peeps outside the room and around the corner, then she goes to the window. She saw Silva-Tongue standing at the edge of the alley talking to Freddy Ready, the drug dealer. She quickly ran back up the stairs and made her exit. She makes it to the other building much faster than when she first arrived. Then Foxy-Roxy left from the building, slipping right by Silva-Tongue and Freddy-Ready like a thief in the night.

Silva-Tongue and Freddy Ready got in his car. Foxy-Roxy was gone into the night holding two duffel bags. One filled with money; the other filled with ten keys of cocaine.

She then ran across the street to a city park. Then she ran into the parking lot where her vehicle was. She threw both duffel bags into the back seat and drove off.

Foxy-Roxy calls Nicole Lovie (Elise) and says, "Operation *birdcage* in progress!"

Nicole Love (Elise) replies, "Roger that! Next operation sleeping wolf is go." Another unidentified voice came on the audio-com also replies, "Roger that!" They all hang up; quickly they go back to radio silence.

CHAPTER 25

The Barbeque

Friday night, at 7:15 p.m., Nicole Love (Elise) and Mr. Andre finally pull up to Mr. Remy's residence. He was standing outside in his driveway, waiting to greet Nicole and Mr. Andre.

Mr. Remy is an Irish American, with salt-and-pepper colored hair cut clean in a military style. With a healthy grade of hair that covered his face looking like Grizzley Adam, a character on a TV series back in the '70s. He was a very fit sixty-five-year-old man. He had military written all over him. You could tell he was the "don't fuck with" type of man. He stood about six feet, three inches and wore black swimming trunks with no shirt. He was 220 pounds of pure muscle. He had on flip-flops like he was having a swimming party in the backyard, which he was. He was cut up with a much defined six-packs and sported a tattoo on his chest of an American eagle and two machine guns that crisscrossed just under the eagle. There was a skull slightly at the bottom; under the eagle, there were numbers written across his stomach which read *The 121st Regiment US Army.*

To Nicole, she could tell he was a very gentle, tender, and loyal individual to his close circle of friends and family. You could tell that he took care of himself and he was proud of his tattoos and what they represented. Mr. Andre noticed Nicole looking at his tattoos and he added that he was a retired colonel of thirty-five years in the service as a commando in Operation desert storm and every other war after that. The bad asses of the army.

Then Mr. Remy hollered out with a happy roaring scream like a frat-boy in college toward Mr. Andre, and he did the same.

Nicole thought to herself, *My goodness, men acting like men.* Soon after their greet, they tackled each other in a bear hug right in the driveway. It was in a kind of way refreshing to see; they had nothing less but love for each other. Then they both released, looking at each other saying, "All right, all right. Boy, oh boy!" like two schoolboys rehearsing a chant trying to look cool.

Mr. Remy says, "It's been a long time, brother. The kids miss you. They're always asking when Uncle Andre is coming over."

"You know I've been busy at work," replies Mr. Andre.

"I know, I know but you know kids. I was running out of excuses, brother. I am glad you could make it. No matter what the reason. Who do we have here?" replies Mr. Remy.

Mr. Andre introduces Nicole to Mr. Remy.

Nicole says, "Hello, Mr. Remy."

He replies quickly, "There is no Mr. around here. Not to friends to my best friend who saved my life once, but that is a story for another time. Just Remy will do."

Nicole complies by shaking her head in agreement with a chuck-ling laugh.

Remy continues, "Well, before we go inside, I know why you are here, but I have one rule. Relax first, wine down, then we can talk business. But first let us go inside so that I can introduce you to the rest of my family and friends."

They walk inside the house. There waiting in anticipation were Mr. Remy's grandkids. Standing in the living room, dripping water, and waiting on his company to come inside. Remy says to Mr. Andre, "See what I'm talking about. They can't wait till you even make it to the backyard!"

Then two of his grandsons, Billy and Timothy shouted out, "Hey, Uncle Andre and ma'am."

Remy quickly tells his grandsons to go back outside to the backyard. He looks at Mr. Andre and says, "See Andre, you got my grandkids soaking up my living room floor, trying to get at their Uncle Andre. See I told you they missed you!"

Andre replies with a huge smile on his face, "Well, you know I miss them too."

Remy says, "Don't tell me. Go tell them!" Then Remy puts his hand out in a gentlemanly manner showing Nicole the way through his house to his backyard. As they walked to the rear of his house toward the back door, he added by saying, "If you guys want anything to drink, I have a cooler out back full of beer, water, and sodas. But if you want something stronger, the stiffer drinks are in the bar in my study. Andre, you know where it is."

"I know where everything is," Andre replies.

"I have steak, burgers, and hot dogs. I'll just put on the pit if you guys are hungry," Remy says.

Nicole replies, "Well, maybe a beer for now. I'm not quite hungry yet."

"A beer it is. I'll be right back; you guys find a seat," says Remy.

Nicole (Elise) could not believe how big the yard was in the back; it was like a scene from *Little House on the Prairie*. It was about twenty people at the bar-b-que. Mr. Remy had a thirty-year-old son, Chad Remy, and his wife, Clara. They have two sons named Tim and Billy. Then you have Mr. Remy's twenty-eight-year-old daughter named Melissa, her husband, Robin, and they have two daughters, Eva and Erin, and one son, Eric. There were a slew of other family members and a few friends. One by one, they all introduced themselves to Nicole.

Everyone knew Mr. Andre; they all called him Unc. After all the formalities were out of the way. The kids bum-rushed Uncle Andre and pushed him in the pool. As he came up for air, almost instantly he turned into a whole other person. He started laughing, giggling, and playing with the kids as if, in that moment, nothing in the world mattered to him but making the kids and himself enjoy life to the fullest. Nicole saw a beautiful side of Mr. Andre that very few people ever see.

She became engulfed in an emotional state of happiness and relief, and because of that feeling only brought more fuel to her fire of revenge. Then she snapped back and walked over to Mr. Remy and says, "I really am enjoying your family."

Remy replied, "I understand. Well, it was worth a try. Ok. Follow me inside. Let us go talk."

Then he yells to Mr. Andre, "Hey, Bub, we'll be back."

Mr. Andre replies, "No problem. I'm kind of busy here, anyway." Referring to playing with Remy's kids and grandkids. They did not even notice Remy and Nicole exiting and going back inside of the house.

Nicole says, "I don't mean to pry but where is your wife?"

"Mr. Andre didn't mention that?" asks Remy.

Nicole replied, "No. I guess he figured you'd tell me."

"Well, she passed away from lung cancer some years back," replied Remy.

Nicole (Elise) says, "I'm so sorry. I didn't mean."

Mr. Remy replied, "That's ok. I am fine. It has been ten years." Trying to connect with Mr. Remy, Nicole (Elise) mentions her loss, "Well, I lost my mother recently," says Nicole (Elise). Mr. Remy replies, "I'm sorry," speaking so very softly.

Nicole (Elise) quietly shakes her head. Whispers in pain, "Thank you." Then she quickly gets back on point. Talking about the task at hand and says, "I would like to know if all of your guys are trustworthy, as I'm sure you already know we've been having a leaking problem at the agency. Don't know whom to trust?"

He replied, "Yes, I've heard of the complication the agency has been having and that's a shame, but the men I have had under my command I trust with my life, all seventy-five of them are ready when the time comes. We will all be in place and I will know what to do. All you need to do is let me know when our phase comes up." Nicole replies, "No problem." Mr. Remy says, "Ok, now can we please get back to the bar-b-que?" Nicole smiles and replies, "Hell, yes. Let us do this. I'm starving now."

They head out to the backyard and join the group of family and friends. Nicole (Elise) finally lets her hair down sort of speak, she had a blast, and she deserved every bit of it.

CHAPTER 26

Pull the Wool

7:27 p.m. Friday night, Foxy-Roxy calls Detective Skaggs and Rachel. Rachel picks up the call and says, "Foxy, what's your twenty?"

Foxy-Roxy replies, "I'm at the foxhole."

Detective Skaggs overhears her on the intercom and says, "Roger that. Silva-Tongue and Freddy just pulled off headed to the bar-b-que shack. We'll be there in five minutes for the bird feed."

"Roger that, and I'm out," replies Foxy-Roxy. She stays in her car and waits for Detective Skaggs and Rachel to arrive. While waiting in her car, she notices and thinks to herself, *This muthafuckin place is nice. From the outside of the building, blinded with the neighboring structures. Looked fascinating on the outside. It had an old-fashioned Victorian feel of luxury.*

Then Foxy-Roxy without thinking blurted out, talking to herself, so she thought not seeing the residents that were walking by. An old white man and lady, obviously they were a couple. They were both in their eighties; they moved around well for their age. "This is some real low-key double 007 ass shit here!"

The old lady says to Foxy-Roxy in a sudden outburst. Obviously offended, "This is a respectable community. Language like that will not be tolerated around here."

Foxy-Roxy looked around, surprised while showing a puzzled facial expression surprisingly she says, *"What?"*

It was not until then that she noticed the elderly couple that was walking by. They had a little white female poodle that was sniffing all around the car and took a pee on her driver's side rear tire. Foxy-Roxy was livid but still apologized for the lack of respect.

They did not even respond or even care about what the dog had just done to Foxy's car. They glanced at Foxy Roxy and kept walking with their noses in the air. When they got far enough past her, she whispered softly to herself, *Well, fuck you too!*

By that time Detective Skaggs and Rachel had arrived. They pull up on the side of Foxy-Roxy's car. Then Rachel rolls down the window to speak to Foxy-Roxy.

Foxy-Roxy gets out of her car and climbs into the van. Rachel moves to the rear of the van giving her the front passenger seat.

Detective Skaggs says, "I did a whole layout of this place. I noticed there are no back doors. Just a freight elevator that is controlled by the front desk. Which is monitored by an armed security guard twenty-four hours a day. The plan is a distraction tactic for this part of the mission."

"I go in and take him out?" asks Foxy-Roxy.

"No. We have to get past him without him noticing us and stay clear of all cameras. They are everywhere," replies Detective Skaggs.

Rachel chimes in, "That's impossible."

"Sweetie, the CIA is in the business of doing the impossible," replied Foxy-Roxy.

Detective Skaggs interrupts, "I have a plan. Hopefully, it will work if everyone here can act their part to precision."

Foxy-Roxy and Rachel look at Detective Skaggs and ask, "What's your plan?"

"I need you, Rachel, to go inside and seduce the guard. Get him totally into you causing him to be distracted long enough to get us in and out within five minutes. Think you can get him inside the bathroom and keep him there for at least five minutes?"

Rachel, being confident in her erotic abilities, replies, "No problem." She looks at the monitor and checks out the guard and says, "Piece of cake."

"Ok, go and get into position. I'll tell you when to go in," replies Detective Skaggs.

Then he says to Foxy-Roxy, "I need you to do your ninja thing and distort the cameras, and I will get the drugs and money and get into position." Everyone got in motion as Detective Skaggs was retrieving the money and drugs. While checking the bags he noticed the weight was about $250,000 cash and at least ten keys of cocaine from his experience of working in the field. While picking up the bags he thinks to himself, *How in the hell did she get this money and drugs up and down that wall so fast? As heavy as the bags were, even after the nano surgery. His body was still tender from that ass-whomping that Momma King's henchman inflicted upon him.* He grunted and held his pain face inward definitely, did not want the girls to think he could not pull his own weight sort of speak. He grabbed the bags in pain, throwing over his back. Then gives the order for Rachel to commence with operation, "Pull the Wool." Then Foxy Roxy pulls out a small round silver mechanism with a black button and pointed it toward all the security cameras temporarily disrupting their function up to five minutes. Clearly some deep CIA component that has never reached the public eye until now.

Rachel walks inside the building, all sexy like. Licking her lips at the security guard. She gets to the front desk and begins provocatively leaning over the top of the front desk, with her breasts almost coming out of her blouse. She did what she had set out to do. Very quickly.

The guard was a five foot, ten inches young, slightly overweight, Caucasian man about twenty-six years old. His name tag says Barnes. Just one look at him, you would think he was a sheltered child. Growing up with no friends exuding a nerdy geek like personality. Looking like a prime candidate of a person that visits Star Trek conventions every year. Racheal was able to read all that with one glance of his face as she walked in; that is how coldblooded she was with her skillset.

Racheal says, "Hello, baby. How are you doing today, sugar?"

Mr. Barnes musters up a nervous, "Hello. I am fine, ma'am. How can I help you this fine evening?" Racheal quickly responds by saying, "Look at you, all respectful and proper like." He immediately blushes. Racheal thinks to herself got him. Then she gives Mr. Barnes a fake name, "My name is Naomi Brown. I'm here from out of town visiting my cousin Shawn Brown." Then Mr. Barnes, after recognizing the name, jumps up to his feet and says, "Oh sorry, ma'am. I didn't know." He apologizes over and over again.

Finally, Rachel (Naomi) interrupts and says, "No need for that. I won't say anything if you won't."

Mr. Barnes smiles and says, "Well, Miss Brown, your cousin Mr. Brown is not presently inside of his residence, and I can't let anyone inside without prior approval. I hope you understand my dilemma, ma'am."

"Yes, I understand. Well, do you have a place where I can wait?" asks Naomi.

"There are some chairs at the front of the building as I'm sure you already seen entering the building. If you would like to wait there will be ok," Mr. Barnes replies.

"Well, to me that's a little too far. I need somewhere a little bit more intimate. If you catch my drift," Naomi says with a soft grin to the right of her mouth.

In an anxious and slightly nervous way, he quickly added, "There is a break room down the end of the hall that is used for security personnel. I have the key and would have to let you in. Do you think that would be sufficient?" asks Mr. Barnes.

"If you would be so kind, that would be great, thank you. You are such a gentleman. I can wait there until my cousin shows up," Naomi says. Thinking to herself, *Perfect.*

"Well, I am not supposed to leave my post," says Mr. Barnes.

"Please, sweetie. Please do it for little ole me. I will certainly, and I mean certainly will appreciate it." Naomi says as she grabs her breasts and runs her hands all the way down the contour of her body as to get a rise out of Mr. Barnes. It worked.

Mr. Barnes replies quickly, "Hold on one minute. I think I can help you with that." He opens the top left drawer of his desk and pulls out a sign which read, "Back in twenty minutes." He placed the sign on the front doors of the building, then locked the doors. Then asked Naomi to follow him.

Once Detective Skaggs and Foxy-Roxy watched him lock the doors, they prepare to enter the building.

As Naomi walks down the hall with Mr. Barnes to the break room, Naomi looks around to make sure no one sees them. Once they get to the door of the break room, he unlocks the door. They walk in as he closes the door behind them.

Naomi looks around the break room and sees two long cafeteria tables with three chairs at each table. There was a microwave and mini fridge on one counter which was closest to the door. There was a coffee pot on another counter which was on the opposite side of the room facing the fridge and the sink. There were two windows one looking out toward the city streets and the other looking out toward the residents' courtyard; both were covered with white mini blinds. There was a half-full water cooler next to the window looking out toward the residents' courtyard. The cup holder on the water cooler was empty. The walls were white, and the floors had white-and-blue checkered tiles very tasteful. There was a first-aid poster on the wall along with other miscellaneous posters of animals.

Naomi thought to herself, Boring.

After checking out the room, she began to rub herself as if she were fixing her skirt. Then she lifts it up slightly but enough to show him part of her pink thong.

Mr. Barnes pretended not to notice but in that moment Racheal a.k.a. Naomi catches him looking at her thong, then he turns around and locks the break room door.

She walks up to him and kisses him softly on the cheek. Mr. Barnes started tearing off his gun belt, then started shedding off his shirt, unbuckling his belt and pulls down pants. Naomi kissed him again and again, then he lifted her up off her feet, putting her on top of one of the break tables. As he lifts up her dress, sliding her thong over to one side, he preceded to pull her leg outward. While caressing her thigh as he entered her. They began having hot, steamy sex.

In the meantime, Detective Skaggs and Foxy-Roxy were in the front of the building. Foxy-Roxy picked the lock with her pocket tools she kept on her person all the time.

Foxy-Roxy says to Detective Skaggs while laughing to herself, "You never know when a door might need breaking and entering."

Detective Skaggs chuckles softly as they enter the building. They were down to four minutes and thirty seconds. Det. Skaggs tells Foxy-Roxy, "Get all of the cameras that recorded us and wipe them clean with this," then he gives her a black handle-looking thing with a flasher head attached at the top. The gadget could erase and automatically set an elapsed time setting by clicking the mechanism once around the handle. Some super high-tech shit. He had on his person. He continues to say, "Do you know how to use this?" Foxy replies, "Hell, yeah. I read about it I never thought I would see one so soon." Skaggs says, "You're right. They're not out yet. It is the prototype, desperate times desperate measures. I need you to access his computer, initiate a 60-second delay to have enough time to steer clear of the building."

"Roger that. I'm on it," replies Foxy-Roxy.

Then Detective Skaggs gets access to the top floor where Shawn Brown, the Pimp's residence is located. Det. Skaggs thinks to himself room number 69 … hmmm, an appropriate number for such a man. He then gains access to the elevator and gets to the front door of Shawn Brown, the Pimp's condo.

He did not say anything to Foxy-Roxy when she made a comment about her breaking and entering abilities. His skills were just as good, then he does his thing.

He gained access to the condo. Quickly and quietly deposited all the drugs and money inside on the living room floor. He closes the door locking it back. The countdown is at two minutes and twenty seconds and counting. Skaggs hurried back to the elevator, calling Foxy-Roxy on the watch com.

Foxy-Roxy responds, "Waiting on you."

Then Detective Skaggs calls Naomi (Rachel) on her earpiece. She answers, but her tone was different. There was heavy breathing, sexual groans, huffing and puffing. Making a hell of a lot of moving and bumping noises as well.

Detective Skaggs calls her name again saying, "Rachel … aww SHIT!" Thinking to himself, Is she fucking this guy? He says to himself, in disbelief. Then speaks to Naomi again, hoping to hear a response. Finally, she responds to Skaggs thru her earpiece. Skaggs says, "I can't believe you. You were supposed to act like a hoe! Not be a hoe! I can't take no more of this!" He tells her with an irritated voice, "Hurry up! I do not mean get a nut. I mean the mission is finished. Hurry your ass on. We are leaving."

Foxy-Roxy is listening to everything that is being said. She was laughing so hard that she fell out of the computer chair she was sitting on! She jumped up to her feet when she heard Detective Skaggs coming from the elevator.

He makes it back to the front desk, then Foxy-Roxy started the 60-second delay, then they exit the building.

Naomi stops and gets up!

Mr. Barnes screaming loudly, "WHAT ARE YOU DOING? I'm not finished!"

With no regard to what he was yelling to her, Naomi pulls her skirt down, then exits the break room, running toward the front door and out of the building.

Mr. Barnes tries to go after her to see why she was running. He forgot he had his pants down to his ankles. He tripped and fell down hitting his head on a chair, then fell onto the floor knocking himself out cold.

Naomi (Rachel) makes it back to the van. Both Foxy-Roxy and Detective Skaggs are looking at her crazy as hell. Detective Skaggs drives off saying, "Mission accomplished."

He then turns back and looks Rachel directly in her face.

Rachel looks at him, "What! Why are you staring at me like that?"

"You know why? Yo ass got issues. You were supposed to seduce him. Not really se-duce him!" Detective Skaggs replies.

Rachel was not even slightly embarrassed. Racheal says, "What difference does it make? I got the job done, didn't I? Plus, I needed some. I was in a drought for a good while and he was convenient." Foxy sitting there tickled to death replying, "Shit, I don't blame you, girl. Get it where you can get it; that's all I'm saying." Det. Skaggs says, "Please, Foxy, don't encourage her." Then Skaggs comments on Racheal's last statement replying, "You certainly did, and you a freak. You know that, huh! Do you? Super freak!" he asks.

Jokingly Rachel says, "I did it for my country, anyway. I'm the one that should be mad. Shit, I didn't even get a nut." Then Foxy Roxy and Racheal looked over at each other and bursted out laughing at the same time.

A disgusted Detective Skaggs just shook his head and says, "Whatever."

"Are you jealous?" Rachel asks.

"Not on your best day," Detective Skaggs replies.

Then Foxy-Roxy chimes in on Rachel and Detective Skaggs saying, "Ok guys, enough is enough. You two fight like an old married couple."

"Yeah, whatever." Rachel and Detective Skaggs say almost simultaneously.

Detective Skaggs shoots the conversation down by saying, "Ok, guys. Let us get in touch with Elise and let her know everything is in place." Leaving Mr. Remy's house, the cell phone rings. Mr. Andre answers Nicole's (Elise) phone. Detective Skaggs could hear a drunken slur in the background and recognizes the voice. It was Elise, and she was drunk!

He felt a little relief that she was feeling somewhat better than she was earlier this week, considering all the horrific events. He thought to himself, Good for her, she needed to release some stress. Due to the loss of her mother and not knowing where her little sister is. He thought to himself, I hope she goes *and gets some well-deserved sleep*. "Take her home to lay down. She needs to get some rest. She has a big day ahead of her tomorrow." Detective Skaggs says to Mr. Andre.

"Ok, I will make sure she is taken care of. Don't worry," Mr. Andre replies.

Mr. Andre hands Elise the phone.

Foxy-Roxy and Rachel both tell Elise, "See ya tomorrow, girl. Get some rest."

Elise musters the energy and replies with a drunken slow slur, "I love you, guys. Good night." Then passes out in mid-sentence.

Mr. Andre picks up the phone and hangs up.

Foxy-Roxy asks Detective Skaggs, "Are you going to call and check in with the headquarters?" Det. Skaggs says, "Yes, I am. You and Racheal go together. I'll meet up with you guys later, ok."

Foxy-Roxy and Rachel both agree as they exit the van. They get into Foxy-Roxy's car and drive off on their way to the bar-b-que shack.

Detective Skaggs drives to a grocery store parking lot and sets up. Parks the van, then goes to the rear of the van and started to contact Chief Williams to brief him on the status of *Operation birdcage and sleeping wolf.*

CHAPTER 27

The Merchandise

7:55 p.m. Friday night, back at the Shadow Lady's (Momma King) lair, inside the fiery-red colored ballroom. There sits Momma King at the head of the table. There on the other end of the table sits Shawn Brown *the Pimp*.

Momma King says, "I've been getting calls from some of our high-end clients about merchandise they have not received yet. You know that the funds that I get from these individuals pay for all my work and pleasures! If anything jeopardizes that we have a big ass problem. What the hell is going on with Freddy Ready?"

Shawn, the Pimp, replies, "I've been trying to reach him on his cell. For some reason, the message from his phone keeps repeating, out of the area, and the call is not possible; please try your call again later. I called his stash house and there was no answer there either. I even called his clubhouse; no one has seen him. I have been trying his phone for the last forty-five minutes. I will get to the bottom of this problem soon as possible."

"You do that. I'd hate to find out that you couldn't handle your little street thugs," Momma King says.

"Yes ma'am. I understand," Shawn, the Pimp, replies nervously.

The phone rings and Momma King picks up.

On the other end of the phone was Darius Khan. He says, "Mother, I'm in position. I have alerted our spies inside the CIA building and they are in position. Just waiting on your word to execute the plan."

"Good boy, Darius. Well, what about those meddling, half-witted girls, that have been snooping around? And that worrisome Skaggs?" Momma King asks.

"I have men on them as we speak. I think something is going on. They were seen coming from the stash house, also coming from Shawn Brown, the Pimp's residence," Darius says.

"So now I know why my merchandise has not been delivered! I got Shawn on that issue. I need you to retrieve my other merchandise sort of speak, and get your men to bring those bitches to me and kill Detective Skaggs and try to get it right this time. Do you understand me?" Momma King says with anger and frustration in her voice.

"Yes, yes, Mother. I'm on it now," Darius replies.

Momma King hangs up the phone.

Momma King says to Shawn Brown, "Those bitches were seen leaving Freddy Ready's stash house and your residence! Is there something you need to tell me?"

"Tell you what? No! I do not know why you are looking at me like that. Hell, I am just as surprised as you are. But I'll find out!" the Pimp replies angrily because he does not know what is going on! This bastard will not answer the phone and now this bullcrap.

Shawn Brown gets up from the table. Honestly trying to get away from Momma King. He knows her and how she is then he says, "I promise you immediate results."

Rolling her eyes at Shawn, she commands two Red Scorpion guards to assist him.

In the back of Shawn's mind, he knows she loves him. Knowing that he knew she would still kill him in the blink of an eye. Realizing this he also knows failure is not an option for him. When they leave, the first thing on his mind is Freddy Ready. Shawn Brown and his cohorts make a mad dash to his residence.

CHAPTER 28

The Abduction

8:00 p.m. Friday night, back at the CIA Headquarters. All of a sudden, a bomb goes off inside the building! Then another bomb goes off underground. Causing a back draft of force to shoot all the sewer lids within a ten-block radius; they shot about fifty feet in the air. Which causes havoc in the city. Fire trucks, police, and ambulances are all responding to the bombardment of calls all over the city within a ten-block radius of the CIA building. EMT services and law enforcement are running themselves thin trying to restore order in the city. There was a major need for medical aid for citizens affected in the city that had been hurt by the bombs. This terrible act was only a distraction for a more sinister agenda.

Coming out of the front doors of the CIA building, gas masks on. An array of tear gas cannisters were thrown from the hands of the twenty-five members of the Red Scorpion Clan accompanied by their fearless leader Darius Khan. Being dragged from the building was Chief Williams. He was badly beaten in and out of consciousness with his hands tied behind his back completely subdued.

Immediately they were met with resistance from the Houston Police Department, other local authorities, and even the Texas Rangers in a matter of minutes. They were behind the vehicles talking through a foghorn. Telling Darius to surrender and release the hostage.

Darius stopped, looked out at his surroundings. There were twenty police cars, twelve Texas Rangers vehicles, and a SWAT truck. All the guns were pointed right at him and his twenty-five men. Darius speaks into his watch and says, "Now."

Then there was a loud echoing sound. The ground was vibrating. Then a gust of air started blowing hard. The police were trying hard to recover from the sudden surprise of wind. By then they looked up into the sky. There flying were three black-and-red Apache helicopters armed to the tee. Then a voice came out of an intercom, giving the police a command saying, "Drop your weapons and back away!"

But of course, they did not listen. They tried to apprehend Darius and his minions by running across the grounds of the CIA building. There were about fourteen police officers coming right at him. As his twenty-five cohorts were protecting the rear and right and left flank. As they had an edge over the cops, they were completely covered from head to toe in a specially made bulletproof material, ten times the strength of a normal vest and five times a SWAT vest. Then, SWAT started shooting at two of the helicopters that were in the air. The third helicopter had descended in front of the CIA building. They were in the process of trying to retrieve Darius and Chief Williams.

The Red Scorpion Clan started advancing toward the police that were running toward them and trying to subdue them. SWAT started taking shots at the Red Clan. The two Apache helicopters were in the air just above the police and the Red Scorpion Clan. One unloaded fifty caliber rounds on the front line of the police vehicles. Glass and hot metal was raining down on the cops like acid rain. Completely immobilizing most of the police. The second helicopter landed. In a desperate attempt to save the chief, the police made a mad dash across the lawn of the CIA building. Then the Red Scorpion Clan started taking the police out that headed toward them by throwing stars (small knives shaped like scorpions). It was unequivocal and utter chaos both sides fighting to the death. The only ones that were dying were the outmatched police. They still would not waiver even in the face of inevitable doom.

While the fighting continues between the uniformed police, SWAT, and the Red Scorpion Clan, Darius puts Chief Williams inside the helicopter yells to the pilot, "WAIT!" Shaking his head up and down.

"Ok," the pilot replies.

Darius turns around and almost immediately he is met by eight police officers completely surrounding him. Running toward him with their weapons drawn! Darius pretends to surrender and he gets down on his knees. As the police get closer, within five feet. He jumped to his feet pulling out his special fighting weapon. (It had two silvertip knives and red steel chains that contracted back and forth at his will. That was attached to some sort of Japanese ancient cartridge rumored to be a part of the Ishi-o-him deadly white arts. It was called the weapon of "THE MIGHTY ONES." Back to the action with Darius Khan and the eight policemen.)

Turning a front flip in mid-air. He swings his left arm one way and the right arm another way, slicing off all eight of the officers' hands that held their guns! He did this while in mid-air! Then he landed with both feet on the ground before their hands holding their weapons could hit the ground!

As the policemen fell to the ground screaming in agony and pain, SWAT starts shooting at Darius even more intensely. Then the two helicopters in the air increased their fire power (to the run or die level). Once Darius was safely inside the chopper, they unloaded once more before leaving a barrage of 50-caliber bullets at the police that seemed to go on forever. The police completely outmatched terribly. The time it took the national guard to get there. They were nowhere to be found. Their satellites were no match for the Shadow organization; her tech was decades past current tech. She had all the right people for the right price, not to mention millions of loyalists. That believed in her 'cause even though the ones on the outside and the inside vaguely knew what that was. The rain from hell was finally over, as Darius's helicopter disappeared out of sight with the chief lying down on the floor at his feet.

CHAPTER 29

The Crash

Friday night at 8:03 p.m., there sitting across the street from Detective Skaggs were two Red Scorpion Clan's men. That came in contact with Det. Skaggs as he was leaving Freddie Ready's residence. They had no prior knowledge of what had just transpired. One was hanging up the phone after receiving their orders. The other one was just sitting in the passenger seat of the car preparing for his orders. They were in a black Mustang. The driver starts up the car and both put on their seat belts and helmets. Then the one in the passenger seat opens the glove box and pulls out two mouthpieces. You would think that by this action they must do this kind of thing all the time. Now they are ready; both men try to motivate the other before the job by savagely hitting their selves and each other. Seemed like some sort of ritual. They both pull their AR-15 assault rifles cocked and ready with a banana clip from behind their seats. They look at each other once more, nodding their heads up and down. Letting the other one know they are ready.

They are across the street from the Kroger grocery store, where Detective Skaggs is parked trying to make a top secret phone call to the CIA Headquarters, but for some reason, he becomes puzzled in why he can't get through. The driver puts the car in drive and hits the gas pedal and revs up the engine! First gear. Second gear! As his speed increases to 35 mph, he shifts into third gear! He throws it into fourth gear as his speed increases over 45 mph! Hitting the van on the driver side while it is still parked. Kaboom! The noise came out of nowhere putting that silent night in a gripping choke hold.

Luckily, Detective Skaggs was sitting in the rear of the van. Due to the impact, the damage to the driver's side in disbelief was only slight damaged. Detective Skaggs still aching from the beating Momma King's minion put on him. It was a miracle that he was not dead or severely injured! He was shaken up and got a hell of a lot more bruises, but fortunately for him, nothing serious.

Across the street there were two police officers that were on a coffee break sitting at the donut shop. They noticed the loud banging sound that came from the grocery store parking lot. As people were coming toward the van from everywhere to see if everyone was all right.

Forty-five seconds later, the Scorpions get out of a mangled car. Pull out guns and started shooting in the air to clear the crowd. Not seeing the cops behind them, coming up out of the donut shop that was in the same shopping plaza. Then the Scorpions start shooting inside the van. Trying to make sure Detective Skaggs was dead!

They got off one round before they were ordered to "drop their weapons on the ground!" The Scorpions turned around to shoot the cops! Then the cops dropped them both before they could get one shot off in their direction!

Detective Skaggs heard the cops calling for him saying, "Hello! Is anybody inside? Please answer!" Det. Skaggs moaning and whaling, he replies, "I am trying to muster up enough energy to speak up to you. As you can see, plainly, I am considerably in a lot of pain." "Okay, Mr. Skaggs, just calm down lie still. An ambulance ETA will be here about eight minutes." The officers try to disperse the crowd and maintain order; they started setting up a perimeter as they waited for backup.

An ambulance and fire truck pulls up twelve minutes later in the parking lot. The firemen help Det. Skaggs out of the damaged van. From the rear with the Jaws of Life because the van was jammed shut because of the force of the impact. Then the paramedics help him on the gurney. He then tells them that he is a cop in the middle of an operation that is a matter of national security. "I need a secure line in order to call my boss." Detective Skaggs says to the paramedics, after he shows his badge to the first and second police officers. Seeing that he outranked them, they became more helpful.

The second police officer came up and says, "Det. Skaggs, sir, I checked your vehicle out when those idiots hit the driver side of your van. Your van had a custom-made job done. Layered in a cast of iron bars something stunt drivers use. I know this because my brother-in-law is a stunt driver for Universal Studios. This van must have been used in a movie. They take precautions like that to protect stunt drivers. Your van is not totaled; it hardly has a scratch, just some dents here and there. Your van, sir, is still operational. You just have to change out your front door panels and side rear fender panel that folded in wedging you between your equipment you had in the back of the van. Eighty percent inside is operational. The 20 percent is your totaled equipment that we cut you out of, sorry.

Det. Skaggs replies, "No need for sorry shit; you saved my life."

The EMT says, "Technically, this van and that equipment saved you. You must have a guardian angel sitting in your lap. The inside of your van has not been touched except for the part I just told you about. That is including the engine as well, not a scratch. Do you need for us to take you to the hospital? If no sir, we have to leave.

"We have another emergency downtown." Det. Skaggs replies,

"Thank you, officers, but those guys were trying to rob me. Thank goodness, you guys were in the right spot at the right time. I am ok." Det. Skaggs hoping they bought that bullshit lame ass lie, and what do you know, they did.

The paramedics check his blood pressure, take his temperature, wrap his new bruises up, and let him go.

Detective Skaggs, as quickly as his injured body would allow him, he seen his cracked-up phone under the crushed-up equipment. He gets on the phone to call the ladies. But there was no answer. Then he tries to call Chief Williams, again no answer. Then the director calls him right back informing him that the CIA building has just been terrorized by the Shadow organization. The chief was taken. Agent **888** wanted to come in, but the director ordered him to finish the mission. Reluctantly Agent **888** complies with the director's orders. Then he thought to himself that is what they were referring to downtown. The reason the emergency servicemen left so quickly.

Det. Skaggs, a.k.a. Agent **888**, was very frustrated not being able to get in contact with the girls mostly, the line was busy. Elise was in bed knocked out drunk, so that was a dead end. Still, he gathered himself together and rushed to the bar-b-que shack to join the girls as he continuously tried to reach them to give the bad news.

CHAPTER 30

The Chase

8:05 p.m. Foxy Roxy and Rachel are driving toward the bar-b-que shack to meet up with Silva-Tongue and later Agent 888. Rachel sitting on the passenger side of Foxy-Roxy's car listening to her talk about fashion and makeup tips. Then suddenly she notices a red truck coming up from behind at great speed. Rachel starts to inform Foxy-Roxy on the potential threat coming up on them from behind.

Foxy-Roxy notices it too! She says to Rachel, "I see it, girl. Hold on tight. It's them fucking Red Scorpions. I saw them two minutes ago. I was just trying to get away from the public before they started trying to kill us!"

Rachel replies, "So what do you want me to do?"

"Get the guns loaded and get ready, 'cause it's fixing to be a very bumpy ride, girlfriend!" Foxy-Roxy replies. Foxy had a 2020 blue BMW, she increased her speed.

Then the Scorpions increased their speed, and the adventure begins.

Finally, the girls find an exit away from the public, an old, abandoned fishery company. That had a lot of old roads filled with a lot of tall grass making it difficult to see clearly. They soon get dumped out onto an old farm road filled with red dirt and large gravel rocks in a rural area making it hard to drive on. It was a lot of curves in the road. There were so many trees on both sides of the road. It was very difficult to navigate, not to mention gage an escape plan.

Then suddenly the Scorpions make their move. In less than a second, they came very close to the BMW on the first turn onto the farm road. By the third turn the Scorpions ram the back of Foxy-Roxy's BMW.

That act pissed her smooth off. Foxy-Roxy says to Rachel, "Brace yourself. I'm (ebonics: gonna; means going to try and whatever word or phrase that comes after gonna) kill these muthafuckers! They starting to get on my nerves!" Racheal getting off shots with every chance she gets. While at the same time trying to hold on for dear life.

Foxy Roxy turns down another dirt road, racing through shrubs and bushes. Doing her *Fast and the Furious Tokyo Drift* as she rolls over loose gravel and dodging large rocks and trees.

Rachel takes a 9 mm and starts shooting again at the Scorpion clansmen's front tires. Foxy was like a pro drifting literally across the loose gravel. Gave them the advantage if only for a few seconds. While Foxy was embracing her inner Michael Mario Andretti.

Foxy Roxy says, "Rachel, wait for your shot. You will know it when you see it, girl, trust me. See the bullet, be one with the bullet, enter the bullet. Is it troubling? I am finding time to bullshit when we're on the verge of being murdered. I think because of doing this type of work, I feel has made me a little numb to traditional ways of normalcy." Rachel replies, "GIRL, I am excited too. At times, I feel I have borderline psychopathic tendencies, but this is not a *Dr. Phil* moment, BITCH OKAY! To answer your second question. YES, VERY TROUBLING AND I'M PRETTY SURE I CAN SAY, THIS IS NOT NORMAL, GIRL! But then again shit what is normal? We got a Trump. OH YEAH! GOT MY SHOT. COME ON, FUCKERS, ONE MORE BOUNCE!" The Red Scorpions' truck popped up again and Rachel timed it perfectly. The bullet seemed like it left her chamber in slow motion. Flying through the air missing the trees, the bushes, and the large rocks. The bullet acted as if it had intelligence striking the brake line of the clansmen's truck. In the one place where movements of brake lines are not expected. Rachel replies, "BULLSEYE, BITCH."

"Ok, Rachel. I have a plan," Foxy-Roxy says.

Rachel replies, "I'm sure glad because we are running out of bullets."

With Foxy-Roxy steadily looking behind her at the Scorpions, then she saw a black-and-yellow dead-end sign. That read 1000 feet cliff quarter mile ahead STOP TURN BACK. She runs over the sign quickly before the Scorpions could see it!

She was ten feet from the cliff then she made a hard right! The Scorpions did not see the sign, therefore had no time to respond to jump out. That was her plan! About the time the Scorpions noticed the cliff it was too late. They plunged to their deaths, SCREAMING IN HORROR.

Then Foxy-Roxy brought the car to a sudden stop. They were both breathing heavily. Then Rachel looks at Foxy-Roxy, and Foxy Roxy looks at Rachel.

They say at the same time word for word like they had the same brain, "That was the shit! You should feel my heart right now. Feel it, feel it."

"My adrenaline is off the charts," in the same breath, "Them assholes fucked my blue baby up, damn," Foxy-Roxy says.

Racheal replies "Mine too and yeah, your shit twisted." Foxy says, "That bitch is going to buy me another one, mark my words."

Then the phone rings scaring them both half to death. In addition to that, their watch coms were busted. The secret red phone hidden underneath the dashboard was fully functional. Somehow it managed to slide to the rear of the vehicle without a scratch on it, go figure.

Rachel gets out of the car, goes to the back of the vehicle, looking to find the phone that was ringing. She finds the phone all the way in the rear of the vehicle. She picks it up, then the phone went dead! No charge. They placed the phone on the charger to recharge it. Rachel says, "And the hits just keep on coming."

Foxy-Roxy looks around to see where they were; both did not have a clue.

Rachel says, "Yes to what you are thinking. We are in the middle of nowhere."

"We just have to follow our skid marks. We'll be ok," Foxy-Roxy assures Rachel.

The ladies do make their way back to the main road. By then the phone had a charge long enough to use the GPS to point them in the right direction home. They headed to the bar-b-que shack to meet up with the gang.

CHAPTER 31

Knocking Boots

Later that Friday night, finally making it back to the motel, Mr. Andre walks Nicole Love (Elise) upstairs to her room. He lays her in the bed and takes off her shoes. Then he takes her jacket off. As he rises up to cover her with the blanket, she grabs him by his shirt, and he says, "Girl, what are you doing?"

She replies, "I'm drunk but I know what I'm doing."

"I'm not taking advantage of you, Elise. Yes, you are drunk. That wouldn't be right," Andre says.

"I'm aware of what is going on right now. I need this, Andre. I lost my mother or the woman who I thought was my mother, and the woman that is my new mortal enemy is really my mother! A cold sadistic bitch. Maybe that's who I really am," Elise replies.

Andre replies sympathetically, "Don't you say that. You are nothing like her. We might not be able to pick who we come from or who we are kin too, but we can surely pick what we become. My grandmother once told me when I was fourteen years old. 'Never let someone's opinion be your reality. You control your own fate no one else.' Your mother is who raised you, loved you. Helped you with your homework at night. Nursed you back to health when you were not feeling well. Attended your sporting events cheering you on to victory. She was the one who took you to pick out your prom dress. She was the one that took your picture when you walked across that stage to get your diploma. Showing you how proud she was of you and your accomplishments. That is who your real mother is Elise, without a shadow of a doubt sort of speak. I do not care if she was related to you as your aunt. She is still in my opinion completely and utterly your mother. Case closed!"

Elise started crying and tears fell from her eyes and down her face; instantly, she became sober from the emotions. Then she says, "Thank you so much, Andre. I really needed to hear that Now I really know I want you now, and I mean right now." She looked up at him with desire in her eyes. "I am serious. RIGHT NOW!" she yelled politely. She grabbed Andre and yanked him into the bed.

Then Andre being a man he gave in.

They started kissing passionately while ripping each other's clothes off. All clothes are off! Rolling all around the bed. Then Andre picked Elise up off her feet in a sexually aggressive way. She wraps her legs around his waist. Then he crashes her back up against the wall ever so gently without hurting her.

He then in that moment enters her. She grabs his back and holds him so tight, as she embraces his manhood thrust inside her. Both became lost in their passion for one another breathing uncontrollably. In delight, they made love all night. Andre still standing and holding Elise up against the wall, in the bed, on the floor, on the desk, then back into the bed!

Finally, Elise was on top. She rode him like a stallion in the Kentucky Derby! Two hours later, she had her orgasm, seconds later Andre follows close behind her in complete satisfaction he yells out saying, "Thank you, oh my god." Then Andre delivers a soft and gentle kiss on the left booty cheek of Elise. Then they both embrace each other as they kiss each other on the lips, then fell to the bed totally exhausted. She flopped her head on top of Andre's chest and soon after, they fell asleep.

CHAPTER 32

Bar-b-Que Shack

Meanwhile earlier Friday night unbeknown to Silva. The current events that was happening to more than half of her team. Had been to hell and back with gasoline drawls on. Freddy Ready and she finally make it to the bar-b-que shack. It started raining like cats and dogs. Freddy Ready says, "Damn! Where did this damn rain come from? I can't see a thing!" Freddy finds a parking spot and continues to say, "We got to get out and go look for that asshole Shawn Brown. Come on."

Silva-Tongue replies, "What I tell you about that Dee-Bo shit?"

"I didn't mean it like that, girl, damn. Man, hoes kill me!" Silva thought to herself, Nigga, you have no idea. "Always trying to be sensitive and shit."

Silva-Tongue trying really hard to stay in character because he struck a nerve with her. On the cool, Silva-Tongue says, "It's not me being sensitive. It's me having some respect for myself while in these streets, playa."

Freddy Ready replies, "Whatever, are you coming? Please."

Silva-Tongue says, "Yes. I'm coming."

They jump out of the car and quickly run into the bar-b-que shack. They got drenched as they stood in the entryway both dripping from the rain. Someone walked up and gave each one of them a warm dry towel.

They approached the door and walked inside. Once inside, a petite, short, skinny, yellow lady with freckles and reddish-black hair standing about five feet, two inches. She is about fifty-nine years of age. People call her Mrs. Sammy. She was an aggressive, but in a nice way, and outspoken Creole woman from the boot, that's Louisiana. She was from the southern part of New Orleans.

She was standing in the doorway. Wearing a hat that read, Bubba's Bar-b-que Shack, a red apron with a hand full of promotional flyers. She was passing them out as people walked in. She also passed out balloons to everyone that came in with small children. Mrs. Sammy greeted everyone with a smile and a good old southern style very loud *HELLO!* To whomever came into the door, letting them know that they were welcomed.

Next in line were Freddy Ready and Silva-Tongue. He smiles at Mrs. Sammy; she gives Freddy and Silva-Tongue a flyer. Then Freddy reaches for a balloon.

Mrs. Sammy pulls back and says, "Sorry, sir. These are for the children."

"I have children," Freddy replies.

What did he say that for? The *black* came out of Mrs. Sammy, like hot lava exploding from a volcano. She asks, with some attitude in her question, "Are they here right now, sir?"

"No, they are not," Freddy replies.

"Then no, you can't have one!" Mrs. Sammy says with a fast snap back remark.

Silva-Tongue giggles to herself as she apologizes for Freddy, then pushes him into the restaurant to be seated. They went to the bar waiting for their table to be ready.

Looking at him crazy Silva-Tongue says, "Why are you trying to get that woman riled up? Shit she's Creole for real. She will cuss your ass out and tell you to have a nice day and mean it. You already know that, don't you?"

Freddy Ready says, "Yes, I know but I don't give a fuck. I'm already pissed off!" Silva replies, "Yeah but don't get us kicked out before we even get in. Damn, nigga, think." Freddy says, "Whatever. I bet you I won't leave. Speak for yourself." Silva replies, "You know what, you are acting like a pompous asshole. You do know that too, right." Freddie tickled started laughing, then said sarcastically, "Yes, I do."

Then a big six-foot, nine-inches, black, 320 pounds man walked up on Freddie startling him. For a moment made his ass jump into a run pose. He was wearing a shirt that read, *I am Bubba.*" He was wearing a red apron. "I am Mrs. Sammy's husband, your host for this evening," Bubba says. He scared the hell out of Freddy for a second. Paranoid ass Freddie thought he was coming to pound his head in for disrespecting his wife but he was coming up to welcome and walk them to their table.

They both followed behind Big Bubba. He seated them, then told them your waiter will be here shortly. He thanks them for choosing Bubba's Bar-b-que Shack, then walks off. As he walks away, he turns back and looks at Silva-Tongue. Then he winked his eye at her. At that moment, she knew he was an operative. The fourth phase is about to begin, she thought to herself.

Then their waitress walks up to the table and introduces herself as Wendy and she would be their server for this evening.

"What would you like to drink first, sir?" Wendy asks.

"A beer,' Freddy Ready replies.

Silva-Tongue says, "A Coke."

"Ok, I will get that for you and here is your menu. I will be right back," Wendy says.

"Thank you," replies both Silva-Tongue and Freddy.

Once Wendy walks away, Freddy comments on the remarks of Mrs. Sammy. He says, "Shit I don't think that lady liked me as soon as she saw me."

"What lady? Oh, that lady. Boy, that lady is not thinking about you, out of mind out of sight," Silva-Tongue says.

"You didn't see the way she was looking at me?" asks Freddy.

"Boy, you are paranoid, shit. Chill out and order something," Silva-Tongue says.

Freddy snaps back with a reply, "This is not a date, woman. I'm looking for that pimp."

"I know that but I'm still hungry, big head," in a play gentle like voice says Silva-Tongue.

"Alright man, shit! Order something and not none of that high ass seafood and steaks shit either because I'm not paying for it," he replies.

Silva-Tongue thinks to herself, *Cheap ass bastard feeling no remorse for what was coming to him.* She gets fed up with Freddy Ready. She signals to the operative to go ahead with the next phase of the operation and so it begins.

Suddenly, Freddy Reddy sees someone fitting the description of Shawn Brown, the Pimp, going toward the bathroom. He quickly gets up and tells Silva, "Order him some ribs. I am going to the bathroom. I think I just saw that asshole walk by. Let me go check."

Silva says, "Ok, if you really think it could have been him." Freddie Ready replies, "Seriously, real nigga shit, I never ever seen any other negro wear his high dollar furry shit but that muthafuckers!" Laughing inside to herself. Silva says, "Whatever, then go check it out." Feeling anxiously to rid herself of his trifling ass.

Then she calls Foxy-Roxy on the phone after Freddie is out of sight. There was no answer. She calls Elise, no answer. She started wondering thinking to herself, Where in the hell is everybody? She called Detective Skaggs and there was no answer. Then she really got nervous.

Bubba seen the worry in Silva's eyes. He takes it up on himself to calm her before the show starts. So he walks up to her and says, "Everyone is across the street waiting. There is radio silence, in case there is someone listening in, due to unfortunate events that I am sure you are unaware of, ok."

"Cool, I got it. I was starting to lose it for a second," Silva-Tongue replied.

"I know. I just came from outside. They told me you would probably be tripping out right now. You guys certainly know each other. They look like they've been through something tonight. I'll let them explain that also," Bubba says.

"Like what? Something happened?" Silva asks.

"I'll let them tell you. Ok," Bubba says.

"Ok. I gotcha," Silva replies.

Then she started thinking about Freddy Ready. She could not stand him and what he stood for. Thinking to herself, Muthafucka be killing people with that poison, taking advantage of human beings that reached a low point in their life. She could not stand by and idly allow this monster to exist anymore. When she thought about his fate, it was a large rumble that escaped out her mouth. It was a very loud laugh of vengeance. She quickly covers her mouth looking around. She senses people looking at her. She pretends she is looking at something on her phone.

For a passerby and those who are so damn nosey listening, she says out loud, "Boy, oh boy. That Kevin Hart, he's a riot. Boy, he is so crazy." Everybody starts to shake their head up and down in compliance with her remarks. They smile even though they had not heard the imaginary joke that Silva-Tongue just manifested out of thin air in her head, to keep her cover. Then everyone quietly goes back to their business, eating dinner like nothing never happened.

She started wiping off her forehead in relief. Then she sits back to wait on the fireworks to come from Freddy.

Freddy enters the bathroom. He starts walking down the row of toilet stalls. At the first stall, he notices a fresh pair of white snakeskin Stacy Adams he always sees the Pimp wearing. He heard talking inside the toilet stall. It sounded like him too. For a second, he wondered where his bodyguards were. Then he did not too much care.

Freddy thinks to himself, *Yes, muthafucka, I got your ass now.* He leans up against the bathroom wall just across from the toilet stall. Shawn is supposed to be in. He becomes so anxious to confront Shawn, the Pimp, about recent events, the taking of his drugs and the killing of his men. It was not like he could call the police for support. Freddy became even more enraged just by thinking about it so much; made himself so furious without thinking, right before he could kick the bathroom stall door in. He was interrupted by a short, bald, Oriental man coming inside the bathroom. He nearly ran into him. He was acting like he was waiting on a bathroom stall. Quickly the little man apologizes for the sudden burst into the bathroom.

Freddy, trying to pay attention to Shawn, in case he comes out, quickly accepts his apology.

The man uses the urinal, washes his hands, dries them, then kindly says, "You have a nice night, sir."

Freddy is so irritated by being interrupted, he rudely says, "Ok, bye," to the Oriental man leaving the bathroom. The man quickly leaves Freddie Ready's presence. Then suddenly Freddie goes back to his agenda. He starts hearing more talking. To him it sounded like the Pimp was boasting about today's events. Freddy got even more pissed. Then in that moment he thinks to himself, *Forget this crap, I'm not waiting anymore.* He screams out saying, "I know that is you inside that stall, Shawn Brown."

He kicked open the bathroom stall door. With his gun drawn and pointed it. To his surprise standing right in front of him was a complete total stranger. Freddie standing there with his mouth wide open. Freddy quickly spoke out loud saying, "I'm so sorry, sir. Dammit! I thought you were someone else, man."

The man says, "Clearly I'm not, mister!" Then he started screaming out loud saying, "Please don't hurt me, sir!"

Freddy jumped back in surprise thinking to himself, It was not even a minute earlier this fucker had replied calmly. Then suddenly he started acting hysterical.

Freddy starts to backtrack wondering to himself, How in the hell did I get this wrong? Forgetting all about the gun he was holding in his hands.

Suddenly, an off-duty police officer, who just happened to be a sergeant for the HPD, was eating dinner with his family inside the restaurant with a colleague from work. They overheard the commotion, got up out of their chairs, and went to access the problem inside the restroom. Before they walked inside the bathroom, they announced themselves as police officers.

Freddy heard the word police and started to panic, quickly throwing his gun to the floor. Then screamed out to the policeman saying, "I'm unarmed. Please don't shoot me!"

Then the off-duty officers walk inside the bathroom. They look around the bathroom, then look down on the floor. They both see the gun at the same time; one of the officers says, "Oh wow, what do we have here? Whos' gun does this belong to? " The man that was assaulted in the stall quickly pointed at Freddie. The second officer took the victim out of the restroom to get his statement and to call for backup. The first officer asked Freddie as he picks up the gun from off the floor with a napkin placing it inside a sink away from Freddie. "Do you have a permit for that firearm, sir? Judging from the expression on your face, I'm going to say that's a big fat no. Am I right, sir?" Freddie disgusted with his self slowly replies, "I want my lawyer, sir." By that time, two security guards walked up behind the off-duty police officers moments later to assist in anyway. One of the security guards whispered in the first officer's ear saying, "They're ten minutes out, sir." The first officer replies, "Ok, thank you." Freddie over-hears the security guards talking to the cop.

Freddy Ready flares up by saying to the policeman, "It was a big misunderstanding, officer."

"I'm sure it was, sir. But you have to understand my position. You threatened a citizen with a lethal weapon. Is that gun registered, sir?" The sergeant asks Freddy for the second time.

"Yes sir. It is," Freddy quickly responds.

"Well, you shouldn't have a problem once you explain it to the judge," the sergeant says calmly.

"So, I'm still going to jail, sir, after all that?" Freddy asks.

"Yes, sir, you are. You can explain everything in court. That being said. Can you please turn around and put your hands behind your back?" the sergeant replies strictly.

Freddy was furious but he did not want to make it worse, so he complies and places his arms behind his back.

Then the officer slowly walks Freddie out of the restroom toward the front of the building.

By that time, a police car pulled up and two uniformed policemen get out of the police car. The sergeant takes one of the policemen to the side and explains everything. While the other officer takes possession of Freddy and places him inside the squad car. Then they placed the firearm in a plastic bag, labeled it as evidence, sealed it up, and placed it in the glove box of the police car. Then one of the uniformed cops turns and reads Freddie his rights.

Freddy says, "I need you guys to tow my car to the pound and let me know where it is."

The police gets his information, then calls for a tow at the bar-b-que shack off Texas Avenue. Then takes him downtown to the Harris County Jail for processing.

Silva walks outside after talking to Bubba on a perfectly executed operation. Bubba and his partner, Agent 777, a.k.a. Mrs. Sammy, quickly brief Silva on another mission that was carried out with precise precision and the art of misdirection Operation sleeping wolf was a huge success. Another close-knit group of friends to a member of the group you may know them as the undisclosed operatives. These extraordinary individuals were a close-knit circle of seasoned agents. What we did not tell you is that they are a covert unit trained in biological weapons of counterterrorism sworn to protect the USA with their lives. Ok, enough of the pleasantries. They created a chemical to immobilize their enemy to ensure a flawless strategic complete and successful mission. Before the enemy knew what happened, they were long gone and out of threat's way. This cbx-14 had the ability to halt all brain activity in suspended animation. For example, picture your brain being an actual DVD and the chemical called cbx-14 is the DVD player having the power to pause you, then hitting play again twenty-four hours later. You never know the difference: twenty-four hours would be like ten minutes to the person exposed to the chemical. The symptoms are painful headaches and a severe case of dehydration. The subject is very thirsty once they become conscious again. Anyway, after the briefing on the other operation, Silva thanks the agents and proceeds anxiously to the meet-up spot. She looks out for Foxy-Roxy, Rachel, and Det. Skaggs. Then spots the ladies down the street. She walks to meet them; she notices the 2020 BMW was beat-up, tree limbs sticking out the grill. The whole vehicle was completely muddy, cracked windows. Then Foxy-Roxy says, "Girl, we had a run in with the Red Scorpions!"

Silva says, "I understand but where is Detective Skaggs?"

Before she could finish her statement, he pulls up. His van all beat-up and bent all out of shape, wobbling. The girls were surprised he could even drive it. Silva briefed them on the other operation called Operation sleeping wolf; she informed them that it was a big success. Last phase is Operation set up. Their celebration was short-lived.

Detective Skaggs says, "I was trying to contact you guys."

Rachel replies, "We had a Red Scorpion problem. Looks like you did too."

"Hell yes! Damn fuckers. busted me up some more!" Skaggs says.

They all started giggling.

Skaggs continues, "I got some bad news. I tried to get in touch with Chief Williams and he wasn't answering the phone. Now I know why Director O'Hara contacted me and informed us that Chief Williams was kidnapped by Momma King."

The ladies were in shock!

"What the hell! We need to call Elise!" Racheal replies.

Detective Skaggs, Silva, and Foxy all agreed that Elise has been through enough for now; they let her enjoy what was left of this night.

Skaggs continues to say that the director wanted them to stick to the mission. The ladies went back to their motel rooms to get ready for tomorrow.

While Detective Skaggs went to the police pound and installed a homing device with a motion detector and bugs inside Freddy Reddy's car for the next phase of the operation. Once he is released from jail on bail, they will know exactly where he is.

CHAPTER 33

The Rosie's Motel

Early Saturday morning, the ladies wake up the next morning, trying to figure out who will be the one to tell Elise that Chief Williams was kidnapped. As they went to her room, they noticed Mr. Andre leaving. He looks at everyone and smiles as he passes by saying, "Good morning." Everyone responds in kind replying, "And a good morning to you too," while giving him the thumbs up. A blushing Andre saying, "Stop it, you guys. I'll see you at the meet-up spot. Godspeed." No one had it in them to tell Andre what happened. He was so happy, and Andre never smiles, at least not round the job.

Then Silva knocks on Elise's door. She looks out the door and sees Silva and lets her in. Then Elise sits down on the bed and says, "Are you feeling good about today?"

"I see you are," replies Silva.

Elise started smiling and in a playful remark she says, "Mind your business, bitch." Then she notices something was wrong by the expression on Silva's face.

"What's going on? You seem distant," Elise asks.

"I can say one thing. You really know me," Silva replies.

Elise says, "WHAT! Tell me."

Reluctantly Silva says, "Well, I was trying to wait until Detective Skaggs came to brief you on the latest events from last night."

"Is it that serious, Silva? Anybody died? Is it my sister? Tell me now Silva!" Elise demands to know.

"It is not Kawanna, Elise. I'm sure of that," Silva replies.

As Silva starts to explain exactly what happened, there was a knock at the door. Then a deep voice on the other side of the door saying, "Wake up, sleepy head. It is us, Detective Skaggs, Foxy, and Rachel. I brought hot coffee."

Elise jumps up to put on her housecoat.

While Silva replies to Detective Skaggs, "Hold on, guys; she is getting dressed."

Elise signals for Silva to open the door. Then Detective Skaggs, Foxy, and Rachel come in. As they enter, Foxy saw Elise's expression and knew that Silva was informing her of Chief Williams's situation.

Rachel says to Elise, "You know already?"

"No, I do not. I wish someone would tell me what the hell is going on. Right now! Please! What?"

Detective Skaggs steps closer toward Elise's bed, then says, "Chief Williams was kidnapped by the Shadow organization."

"What! How is that possible?" she yells.

"She had spies on the inside. Director O'Hara informed me last night. We are to stick to the plan. He also said everyone knew the risk going into this operation."

"Let us just hope and pray that he is still alive, and this mission will not be in vain. Ladies, let us finish this and go get our family." The girls stand up and do a prayer with Skaggs.

Elise says, "That being said, is everything in place, guys? And I want to thank you all for allowing me to get some much-needed R&R even though it was only for one night. It was worth the world to me. Getting myself mentally and emotionally together. Now, I am ready for that bitch; let us do this."

Then Silva chimes in, "Don't forget physically too!"

Everyone started laughing.

Detective Skaggs says, "Everything is a go." He pulled out a device and showed it to them. It was a little black box. The size of a waffle with a digital screen.

It showed a blue-and-red background. He then described it as being a locator recorder, whatever that means; he never expounded on any of these devices, letting us know that they were prototypes. It is a super state-of-the-art device that no one on earth knows of its existence, he brags, except for me and a few of my geek techy friends.

Then Detective Skaggs goes on to say, "Elise, how is Remy and that situation?"

"Like butter baby. Everything is a go," she replies.

Then Foxy says, "Ok, we are set. But I have one question. Do you think the calculations are correct on our entry point for getting inside Shawn Brown's establishment? Huh, anyone!" Elise replies, "Sis, let us hope it does. Please guys, let's stay positive. It has to work; if not, my sister is dead." Then Silva-Tongue comes out of left field with real data that could possibly alternate the mission to a 99.5 percent success rate.

Silva rushes out of the bathroom in a hurry. Holding up her hands while finishing up on brushing her teeth. She spits into the bathroom sink. Then gargles with water, wipes her face with a towel, and says, "What I was trying to say is when we all made it back from our assignments last night. While Foxy, Rachel, and Det. Skaggs were going upstairs to their rooms. One of the girls in the motel recognized us from the introduction from Ramos, remember?"

Elise, Foxy, and Rachel all say, "Yes, we remember. Ok. So what?"

Silva continues, "So what? Well, Foxy and Rachel said they were tired and went upstairs. Well, I was not tired. The girl's name is Jenny, and she had a little party in her room with some of the other girls just before and a few guys from the club. I hit it off with them. It was so great that they have invited us to a Mardi-Gras theme party at the club, bitches. It starts at 7:30 p.m. around the time Shawn Brown wakes up."

Elise replies, "Now that's taking care of business, bitch. Ok, now we have an entry point, a sneak attack instead of blasting first and asking questions later. I knew the odds were not in our favor. Now ask me if I gave a fuck."

Then suddenly Detective Skaggs's little black box goes off. He says, "Ok girls, get ready. Freddy Ready is set to bond out around 7:00 p.m.; that is the longest time Harris County can hold him for us. Then he will be let go and on the move again, ladies." Elise replies, "That's more than enough time."

Elise says, "We will be ready tonight. Everyone set your watch coms in 5, 4, 3, 2, set." Rachel and Foxy said we have new codes because we have new watch coms.

Foxy says, "Elise, that's another story for another time, sis. Right now, I want to stay focused for the fourth quarter, okay and getting my little sister back. Cool." Elise replies, "Kool. I feel that, sis. Everyone go get ready for the big night. Game time, bitches, with the exception of you, Det. Skaggs, big dog." She made everyone laugh but inside she was terrified, not for herself but for her sister and the chief.

CHAPTER 34

Jail Bird

The next day at 5 p.m. the officers kept Freddie inside the system long as they could. Then the Harris County Jail, Freddy Ready was sitting in, allowed his lawyer access to him after purposely losing his paperwork while buying time for the authorities to make a case against him. Then time ran out; a guard calls his name to inform him he has a visitor. Sarcastically, Freddy Ready says, "Thanks, it only took fucking nearly forever. It's about time."

The guard walks Freddy to a visiting room. Sitting inside the visiting room was one of Freddy's paid servants with a law degree. The guard sits Freddy inside the public defender's visiting room. There sitting on the other side of the table was a Jewish man; his name is Louis Stein. He is about fifty-five years old balding man, short, salt-and-pepper hair around his head. He is slightly obese; he was wearing black rim bifocal glasses and carried an old leather strap suitcase that looked like he received it from the past generations in his family. He spoke with a very squeaky, snobby voice that annoyed the hell out of Freddy.

Mr. Stein says, "Hello, Mr. Freddy. I know your upset, but I could not post your bail last night and for good reason; the authorities are just putting your charges in the system. The judge was not available to sign off on your bail until moments ago. I do have good news now. I have posted your bail a few minutes ago.

Then Freddy Ready replies, "Man, shut the fuck up! Your dumb ass voice is killing me. Ok. Ok. When do I get the hell out of here?"

"About 7 to 7:30 p.m., sir. We are waiting on the DA to sign off on it. As soon as your papers are complete, you're out, Mr. Freddie sir," Mr. Stein replies.

Freddy says, "Ok then, so everything is in place like I want. Tell everyone to be at the spot on west 28th soon as I hit the streets." Mr. Stein advised Freddie that the Shadow Lady wanted him to find the Pimp, Shawn Brown, and bring him in. He has been missing for a whole twenty-four hours; no one knows where he is, and he has two Red Scorpions with him as well. Freddie cuts Mr. Stein off in mid-stride saying, "We are going hunting. Tell her, I'll find him and when I do, she can have what's left of him when I'm through!"

Mr. Stein replies, "I don't think you should cause any attention to yourself right now, Mr. Freddy."

"DID I ask your ass for your advice? You did your job half ass if you ask me. Now get your annoying ass out my face before I break my foot off in your ass! Do you understand me? You little fucker!" Freddy Ready says.

Mr. Stein replies, "Ok sir. No need for the name calling." Then he quickly hurries out of the visiting room and out of the way of Freddy's rude and mean remarks!

Freddy looked around and saw that he was completely alone, he then started reminiscing about killing Shawn Brown; slowly it brought a smile on his face.

Then a deputy walks into the visiting room to retrieve Freddy to take him back to his cell block. Freddy looks back at the deputy saying, "WHAT! WHAT! MAN! What?"

"Nothing, man. Come on, your papers came in. Let us go back to the cell and get your shit. You're being transferred to the release cell; you are going home," replies the deputy.

"'Bout time!" in a celebratory manner Freddy says.

Around 7:15 p.m. later, Freddy Ready finally gets his paperwork and makes it outside where some of his guys are waiting for him.

Freddy's cohort says, "Good to see you, boss man."

"What now? Take me to get my shit out the pound. Then we go hunting!" Freddy Ready replied angrily.

"Hunting what?" they ask.

"A bitch ass, pimp!" Freddy yells furiously, annoyed by that stupid ass question.

All of Freddie's cohorts started chanting their death chant they always do right before a war in the streets. Cursing uncontrollably, while making physically obscene gestures on the way to get the car. Right in front of the deputy with no regard for anyone or anything.

They make it to the police impound. Freddy gets his car. His cohorts follow him back to the clubhouse on west 28th. Where Freddy meets the rest of his 135 cohorts. They were waiting anxiously for his arrival.

Freddy Ready drove up to the clubhouse and gets out of his car. He starts greeting them on his way into the clubhouse. His second-in-command walks up to him.

His associates close to him call him Jig Saw. A tall 190-pound, six-foot, four-inches, red-skinned, black man with a smooth player-like voice. He was dressed in a blue jean jumper with a black t-shirt and gold teeth and a hoard of gold jewelry on. He was known for wearing Jordan's latest and stylish shoes. He had every pair imaginable in his possession and always had two chrome pistols under both left and right arms. He always had a fresh razor cut fade with just a touch of curly hair hanging down in the front his head.

"What's up, boss? I got some info on Shawn Brown, the Pimp, for you. Word is he was seen driving up to his house twenty-four hours ago. Checked this out, boss. The muthafucka is still laid up over there; to me, boss, this feels like a fuck nigga job," Jig Saw says. Freddie Ready set on pause for a second, then completely agreed with his second-in-command. Jig Saw asks, "So what do you want to do, boss?"

"Shit get the guys suited up! It's time to talk to that muthafucka!" Freddy replies.

Jig Saw says, "Boss. Are you sure he hit us? He does not strike me as a crash dummy, boss?"

"Well, there is only one way to find out. Let us go ask his ass. RIGHT THE FUCK NOW!"

"You're the boss, and that's why I love the way you think. Gladiators! Let us move out!" Jig Saw says.

CHAPTER 35

Mardi-Gras

Meanwhile, back at Shawn Brown, the Pimp's gentlemen's club, the ladies arrive at 7:40 p.m. in their "hoe-like" attire. Completely blending in with the real hoes. They start laughing loudly and joking with Jenny and her friends that they met at the motel through Gonzo Ramos. As they walk up the stairs, they walk past the guards who were posted at the door and throughout the club.

Silva whispers, "Elise, I told you it would be a piece of cake."

"We are not out of the woods yet. Say, let us go to the bathroom and Silva, keep your Mardi-Gras mask on, girl," replied Foxy-Roxy.

Jokingly Silva replies, "I know what I'm doing, girl."

Elise says, "Ok ladies. Let us get ready. The show is about to start!"

The ladies go into the bathroom and the guards, due to the masks, are totally oblivious to everything that is going on around them.

CHAPTER 36

The Frame-Up

7:30 p.m. Saturday, back with Detective Skaggs and Rachel. They finally pull up around the corner from Freddy Ready's clubhouse on west 28th. Rachel says, "Damn, Detective Skaggs. I mean, Agent 888. Hell, I don't know what to call you shit. Det. Skaggs replies, "Skaggs will suffice. I get that a lot with my closest friends. You know what Racheal, after spending time with you, I see you're a pretty cool camper, and I want you to know I'm really sorry about your family." Racheal appreciating what Skaggs had just said. She gently pats him on his right shoulder saying, "Thank you, man. I needed to hear that we good." Then Racheal gets out of a brand-new van they purchased right before the job covertly to get a closer look. Then Racheal got on her watch com saying to Skaggs, "I wonder what the hell he has going on inside there. Oh shit! Wait, they are leaving!"

Detective Skaggs waves his hand toward Rachel as to say come on back. Then Racheal makes it back to the van very quickly. Then Skaggs hands her an earpiece so she could hear what he was listening to. They did not want to take a chance being detected. Using radio waves, a sure sign of being watched could have been very problematic for the whole entire operation. They both heard Freddy-Ready planning an ambush on Shawn Brown, the Pimp, at his residence. Detective Skaggs and Rachel hurried over to the Pimp's house before Freddy could get there.

They get there five to seven minutes earlier waiting on Freddy Ready and his cohorts. Freddy and his crew finally show up at 7:47 p.m.

Rachel says, "Detective Skaggs, now he's going to kill Shawn if we don't do something."

"Well, that's not my problem. Fuck him! He's done so much bad in this city, and Freddy Ready, I hope they kill each other at the same time!" replied Detective Skaggs.

Rachel says, "Everything must go on as planned, you know that!"

"I will get up high in those trees over there by that fountain and make sure everything does. I'll make sure he stays alive if I have to snipe every one of those motherfuckers, but after that, his ass is on his own," replies Detective Skaggs as he snaps at Rachel, then apologizes.

"I'll alert the ladies …" Rachel says with the shaking of her hand and head in a motion as to say you good I understand, without saying a word.

"Not now, radio silence. No in or out frequencies. This assignment is blind for now. Please stay off the coms, Rachel," Detective Skaggs replies.

"Roger that," Rachel says.

Shawn Brown finds himself lying on the floor next to his babysitters. Not knowing what just happened, he quickly picks up one of the bags. Then orders the Scorpions' guards to get the other bags. As he got to the front doors of the elevators, at 7:46 p.m. he felt drowsy. Thinking to himself then he says,

"Goddamn it! It's a frame-up!" He immediately got paranoid. He did not know whom to trust. Then they rush their way back downstairs to the main lobby.

The security guard immediately walked up to Shawn Brown to warn him about something brewing outside, but Shawn Brown was so angry, he walked up to the security guard and put a bullet right between his eyes. Dropping him like a bad habit, instantly he was dead. Then in a "JOKER" from *Batman*, kind of way Shawn Brown says, "Yeah, what you wanted to say sorry for doing a piss poor ass job. Ok, I accept your apology." He was talking to the man in first and third person as if he were still alive but was not. Then kicked the guard as he stepped over him going out the door to the front entrance of his apartment building. Acting like a true psychopath in every sense of the word. He noticed Freddy Ready's car and a legion of his cohorts right behind him waiting on Shawn Brown to come outside. They were so many constantly pouring into the parking lot. Exiting their vehicles squeezing behind Freddy Ready hoping to be a part of the action.

Then the Pimp says as he walks out to the front of the building, "Freddy, what's going on, brother? I have been trying to get in touch with you for the last couple of hours," then Shawn Brown paused for a second as he was trying to catch his breath. Then he continues to say, "Damn man, my head hurts and I'm fucking thirsty." The Scorpions say the same thing, as usual Shawn Brown pays no attention.

Freddy Ready replies, "Shawn Brown, save that for someone who gives a shit!"

"I know what this may look like, but I didn't take this from you, Freddy. I was framed, buddy. You have to believe me!" The Pimp tries to explain to Freddy Ready. But his plea fell on deaf ears. Freddy says, "If that's true, why did it take you a whole twenty-four hours to tell me? Why not yesterday, muthafucka? You've been here since yesterday. You didn't even try to come get me out of jail." Shawn Brown replies, "What in the fuck are you talking about? I just got here fool." Freddie starts to laugh all crazy like, then he says, "Ok. I'll play along. So tell me what is that you carry on your shoulder and your bodyguard's shoulder?" Freddy asks.

The Pimp tried to explain but it just did not come out right. He turned toward one of his bodyguards like he was going to say something. Then snatched the bag off his shoulder and took off running toward his car. Leaving his bodyguards there to fend for themselves!

Freddy Ready told his cohorts not to harm Shawn Brown, the Pimp, because his ass belonged to him. Then he started chasing after the Pimp, not realizing just how fast he really was! He stopped chasing his ass and started shooting at him so frustrated mad he got dusted in the run!

The Pimp moved like a gazelle, dodging back and forth until he made it to his car! He opens the door, throws the bags onto the front seat, got in his car, and burned off! Tires screeching, wheels burning, and peeling off. All the way out of the side entrance of his residence! Not before Freddy and his minions delivered hundreds of rounds of ammunition all through

Shawn's vehicle, looking like a metal piece of swiss cheese. Being nothing short of a miracle, Shawn Brown still alive and in high speed he makes it to the freeway!

Then Detective Skaggs continues to watch as Freddy Ready and his cohorts enter their vehicles. Peeling off in high speeds! Chasing after Shawn Brown, the Pimp! Running over one of his bodyguards and putting a bullet in the other one's head!

Skaggs gets out of the tree and makes it to the van.

Then Rachel says, "Did you see that muthafucka run? I have never seen any one run that fast! Ever in my life, he could have given Usain Bolt a run for his money!"

Detective Skaggs replied, "Hell, yes! I saw that, shit! Let us go, call the girls, let them know we're on the move!" Racheal says, "Ok. Roger that. Doing it right now." Rachel steps to the rear of the van, as she reached for the sniper rifle out of Skaggs hands, as he gets back into the driver side.

Det. Skaggs starts up the van and they go meet Mr. Remy and his crew to get set up for the big showdown—the final phase of the mission.

CHAPTER 37

The Shadow Lady's Lair

Saturday night 8:15 p.m., the Shadow Lady continues to torment Chief Williams. Trying to get him to tell her the location of the last music boxes. To completely synthesize the white energy injections correctly. Doing this could shift the tables in her favor once and for all. The Shadow Lady says, "Mr. Williams, your resistance is futile. Why do you defy me? I will inevitably find out this information. Why not just tell me now to save yourself the pain and aggravation Well, I must be honest, a part of me hopes you will not comply. I have some new toys I have been itching to use for a while. You think that what you are doing is honorable. No sir, you would be wrong; it is stupid. Trying so hard to protect a cause that does not protect you. How is it that I could walk into a federal building of the CIA in broad daylight and walk right off with your fat donut eating ass. Because you will never have more people more loyal to the agency than I have loyal to money. Everyone has a price; what is yours, chief?

Chief Williams says, "Your head on a platter; make it happen. That is my price, bitch! I am not telling you shit! Lick my ass and part my hair. Do your worst."

She replies, "Really, sir, you don't want that."

As she gestures to Darius to continue beating on Chief Williams who is sitting in the chair half dead. Spitting blood, slowly dripping out of the corner of his mouth. Eyes swollen one more than the other. Scrapes and gashes, bodily fluids flowing up, down, across, and all through his face, down to his body. Mixed with dirty blood, feces, and vomit.

As his beating got worse and worse, he let out a bloodcurdling scream. Then Darius asked him did he have enough. Making another joke after another, the chief knew his time was short. He could feel the life draining right out of him. He made a merciful plea, not for his own life but for the lives of the girls, Elise, Foxy, and Silva; he felt deeply responsible for, on some level. Like a lion that would protect his cubs to the death. Chief spoke with a weak muttering, low scratchy voice, speaking directly to Momma King, a.k.a. the Shadow Lady, "If they would not be harmed in anyway, I will reveal any information that will lead to the discovery of the last of the music boxes."

Anxiously, without a flinch, the Shadow Lady agrees! Then she nods her head toward the chief in compliance. She starts by saying, "Go ahead, Mr. Williams. I said I agree to your terms. I want to preserve them as you do. They are special to me as well. I would not dream of harming them directly. Of course if they are worthy, they will survive. As you already may know or may not know." Looking puzzled by Momma King's last statements, not knowing exactly what she was talking about. The terms were met in agreement. As Chief Williams almost started to tell the Shadow Lady the location of the last of the music boxes, there was an explosion from up above! Inside the Shadow organization's lair above.

Momma King quickly stepped away from the chief and walked over to her monitors.

By that time, Elise, Foxy, and Silva are already deep inside the so-called brothel, hiding in plain sight. Undetected, sitting in a dark corner where there are no cameras. Enjoying the Mardi-Gras festivities. Waiting on the Pimp, Shawn Brown, to show up.

Without warning, the Pimp burst through the doors. Hollering and screaming at the Red Scorpion guards to kill anyone that comes through the doors! Some of Freddie Ready's cohorts without thinking ran in behind him, seconds after him, and they were shot dead along with some Scorpions from outside trying to protect the Pimp, Shawn Brown.

The Pimp scratched his head in disbelief as he looked at the guards as if to say, "Idiots! Not them, you fools. What's coming next!"

Right after that statement, a flurry of bullets followed soon after. Tearing down the whole entire so-called brothel establishment, a.k.a. Shadow organization's hidden lair, now exposed. In a matter of seconds, piece by piece. Many of the Scorpion guards that attended the front entrance and half of the customers caught out in the open were utterly annihilated. Shawn Brown holding the bags quickly ran to the private elevator where the girls were posted up near the hallway to the elevators.

As the ladies saw the Pimp enter the hallway, they allowed him to put in the code to the elevator, then quickly overtook him and his Scorpion guards that were protecting him. They gained access to the secret elevator and stood waiting for the elevator doors to open. Five minutes earlier, as the Shadow Lady walks up to her monitor and sees a flurry of gangsters coming into the club! It was none other than Freddy Ready with an AK-47 shooting up the Pimp, Shawn Brown's Mardi-Gras party.

Then Freddy Ready says, "HELLO, MUTHAFUKERS! Guess who is coming to dinner. I will give you a hint. It is not Sidney Poitier, tender dicks. Thought you could play me for a bitch! No hoes, ALL YOU MUTHAFUCKAS! are going to be my bitches tonight! Trying to play me! Freddy Ready does not get played. I do the playing!"

Then Freddy lets loose a tornado of more bullets that were flying everywhere through the air. Cutting down hoes and guards like they were blades of grass on a lawn.

Suddenly, a hoard of Red Scorpions showed up out of nowhere. Then shiny, silvery stars and sword started flying like projectiles into the air. The Scorpion guards started fighting back in full force! They numbered 305 and up against Freddy Ready's cohorts of 150. Slowly they diminished each other's numbers. All the way down to fifty Scorpion guards and twenty cohorts that belonged to Freddy Ready's gang.

In the meantime, while all the fighting and killing is going on, back at Mr. Remy's meeting place, just outside the club and down the way, there were fifty-five strong and loyal special force soldiers. Detective Skaggs says, "We have confirmation from the ladies. It's time to move in."

Mr. Remy signals to the soldiers to move in. The soldiers went black putting night-vision goggles on. Then they move in releasing flash grenades, completely immobilizing the cohorts and the Red Scorpions' forces before they realized what was going on. As they laid waste to the bad guys, little by little but quite efficiently. They turned the tides very quickly in their favor.

Momma King sees a disturbing turn of events, going on from up above. She signals to Darius to stop torturing the chief and go take care of the mess that was going on upstairs!

Darius says, "Momma King, what about Williams?"

The Shadow Lady replies, "I have something else planned for Mr. Williams. You go take care of those meddling girls upstairs!"

"Yes, mistress. Your will is my command," says Darius Khan.

Then she gestures to Darius in a loving way, like she knew he might not return. Without uttering a single word in some way, he knew. Then the elevator opens and off Darius goes back up top to stop the girls.

The elevator doors open again and there standing in the hallway outside the elevator were the girls with the Pimp, Shawn Brown, all subdued. Foxy says, "Finally get to meet the big bad wolf, so what up trick. I'm just saying you in a hoe house, technically you a trick." Darius became furious about Foxy Roxy's remarks, he replies, "You have a dirty little mouth, little girl. I see I'll have to teach you some manners." Foxy says, "That's what I'm talking about. A man that knows what he wants and goes after it. Well, sugar, come make it sweet." Foxy looks over at the girls and says,

"Take Shawn Brown down with you, guys. I'll handle this situation." Elise replies, "You good, baby girl, for sure? Well, give him hell, then send him there."

Foxy nods her head as the elevator door closes.

While Detective Skaggs storms the rear top of the club. Rachel took out the remaining men at the front gates. Working her way back to the action that was going on inside.

Freddy Ready saw the soldiers that were dressed in all black coming toward him and his cohorts. They tried to keep Mr. Remy and his soldiers away from the hallway where they were posted up at. They were the last of all the men alive. Fighting for dear life, Detective Skaggs successfully takes out the last of the Scorpion guards that were shooting at Rachel from a top balcony near the stairway as she was making her way from the front to the back of the building. Then he sent a portion of his group through the back windows of the building to meet up with the frontal assault group, which Mr. Remy was leading.

Rachel looked upwards through a glass roof portion of the brothel. She saw Detective Skaggs looking back at her throwing her the deuce as to say, "you're ok now." Rachel threw a deuce back as to say "I am good" without saying a word. Quickly Rachel raced to help the ladies, not even knowing if they were dead or alive.

Mr. Remy calls Detective Skaggs on the watch com saying, "Partner, what's your 20?" Detective Skaggs replies, "Me and my guys took out the last of the Scorpion clansmen upstairs. I haven't located Jig Saw and Freddy Ready just yet." Mr. Remy says, "I found them. My guys got them pinned behind a statue. They will not give up; we call them Tony Montanas." Det. Skaggs replies, "Then give them what they want." Mr. Remy says, "Roger that."

Freddy Ready still hiding in the back of a large fifteen feet Roman Centurion statue. He completely eluded Mr. Remy and Detective Skaggs. Then Freddy Ready slowly aimed his 9 mm pistol through a gap from the wall and the statue from a position not easily visible from Mr. Remy and Det. Skaggs's line of sight. He aimed directly at Detective Skaggs's head. His finger was on the trigger. But before he could pull the trigger, Rachel came up from behind and blew his brains out onto the Roman statue, the wall, and up against a white naked lady statue that was close by. It was drenched with Freddy Ready's blood. He fell dead instantly to the floor.

A few minutes earlier, Darius charges Foxy at full speed! While performing a flurry of martial arts' moves! Foxy counters every move! Then he steps back, looks at Foxy, then smiles. Then Darius says, "Little lady, you got some skills. I will end you here today."

Foxy replies, "That's what they all say. You promise, little man. Come here, you so cute. I'm going to fuck you up. Boo! Boo! The only one leaving this fight alive is this fine black ass right here!"

Darius got really pissed. Then he started screaming in a rage and rushed at Foxy again, again and again. This time he did his frontal assault flip while bringing out his slinging chrome blade tips. Which were attached to his red steel cable. Throwing it in a way that was almost unseen to the naked eye!

Foxy countered that move also! Then he did a move that was damn near supernatural. She was barely able to counter that move in surprise. Foxy says, "Ok, you have some skills. This should be interesting." Then he sliced her left arm, not too serious, and kicked her in the stomach. By that move alone you would think the fight was over.

Then she had a flashback and remembered what the director said about the white energy and being born to it. Her body became so stressed that her body went into a primal overload resembling an Amazon warrior; she started to move like Wonder Woman. She felt this super-sensational feeling of power that started to surge through her veins! Her pain went away; her eyes turned into two burning balls of fire.

Foxy says, "What the hell? I FEEL FUCKING GREAT!"

Darius, in disbelief, tried to finish her off by throwing a death blow with both of his chrome knife tips headed straight for Foxy's head; to her, it was like he was moving in slow motion at times.

Foxy quickly grabbed a piece of furniture that got broken up in their fight and threw it toward Darius. His knives got caught up in the piece of furniture. Then Foxy started rolling up the wire getting closer and closer to Darius. Then he tried to do a reversal kick, hitting Foxy in the head. She was dazed for a moment; her power left long enough to bring her to her knees by kicking her in the stomach for the second time. Right before he could deliver another blow on her, Darius's head exploded all over Foxy's face!

He falls to the floor, dead! Then Foxy trying to get Darius all out of her mouth, ears, and hair in disgust, Foxy says, "Thanks, whoever you are but damn you could have shot him in the body. Now I'm going to have pieces of his ass in my hair for days."

Then someone replies, "Well, you're welcome too, Foxy. I see you will never change." Then she feels someone picking her up off the floor gently as she fell in disgust from the nastiness of Darius on her person. It was one of Mr. Remy's guys so she thought. He took his black mask off and it was none other than Mr. Andre.

Foxy, trying to recover from the embarrassment, jokingly says, "What took you so long shit?"

"Sorry, we had our hands full on the roof. I hear Freddy Ready is dead but Jig Saw got away." Mr. Andre replies, then everyone started pouring into the hallway where Foxy and Mr. Andre were talking.

Detective Skaggs, Rachel, and Mr. Remy all greeted Foxy and Mr. Andre at the elevator doors. Then they tried to open the doors to the elevators, but for some reason, the doors were jammed.

Meanwhile, back downstairs in Momma King's lair, the elevator doors open. Elise and Silva walk out the elevator doors with Shawn Brown, the Pimp, right in front of them holding a pistol to the back of his head.

Elise got enraged when she saw Chief Williams sitting in a chair, tied up with wires running all around him. With loads of C-4 strapped to his horrific beaten body. Blood and saliva dripping from the corners of his mouth. Both of his eyes busted and closed shut! He was trying to talk but was hardly making any sense; his tongue was extremely swollen. He kept saying to Elise unintelligently chanting, "KILL THIS BITCH! Kill THIS BITCH NOW. DON'T LET HER GO!"

Elise, then Silva, both turned Shawn Brown around and kicked him to the floor! Then Elise tried to make a move toward Momma King.

Momma King says, "Oh no, no, no, no. Boo, Boo. You want to stay right where you are. I have enough C-4 to blow everyone here and upstairs into the next world, and I will do it, trust me. Including myself, to keep you guys from stopping me! Better BELIEVE THAT SHIT!"

Then Chief Williams started back to tell the girls to, "Kill that bitch, please. Right now. That's an ORDER!"

Elise replied, "Sir. That is one order I can't comply with. We need to get you help, sir. I'm sorry. We'll kill this bitch later; her days are numbered."

Momma King says, "Now, is that a way to talk to your mother? I'm surprised my sister didn't wash your mouth out with soap." Then she smiles.

"WHAT the HELL do you know about being a mother? You might be my biological donor, but BITCH! YOU ARE NOT MY MOTHER!"

"Amen to that!" Silva chimed in.

Shawn Brown, the Pimp, tried to add something, and they all said at the same time, "SHUT THE FUCK UP!"

Then Momma King says, "He's not worth ass that shit comes out of but he's my ass that shit comes out of. Let him go NOW!"

Shawn gets up to his feet and walks toward Momma King. She slaps the shit out of him! Then says, "You are a fuck-up! Get my ride ready."

He nods his head, goes behind a secret door, and disappears out of sight.

Elise says, "I don't understand you. How could you kill your own sister?"

"When I killed her, she was no longer my sister. She betrayed me by working with these assholes! After what they did to me, Foxy and Silva's parents, I asked her to join me many times. She refused!" Momma King replied.

"WHERE IS MY SISTER, Momma King?" Elise asks.

"Your sister! You mean your cousin! I thought you guys hid her from me. I sent people to retrieve her but she was not on the campus at the time. Hell, your guess is as good as mine. I really do not know. There is a lot of shit you do not know, Elise. Now is not the time; maybe another time, sweetheart. If you survive, then you are worthy," Momma King says in a sadistic humorous frustrated voice.

At that time Momma King noticed that the secret elevator was closing behind her, she pulled out a device and warned Elise and Silva saying, "DO NOT follow me or everyone will die!"

Halfway inside the secret elevator doors, Momma King walks through it and throws the device at Elise and says, "I don't want to kill you. There is so much I will let you know, only if you can stop the bomb on your boss and make it out alive. See you later, baby girl."

Elise sarcastically says, "No, I'll see you real soon, Mommy." The secret door closes. Then Silva tries to reopen the doors, but it is useless.

Elise quickly goes over to Chief Williams desperately trying to free him.

The elevator doors open it is the *crew*. Detective Skaggs says, "Where is Momma King?"

"The bitch got away through that door!" Silva says.

Mr. Remy and his men try to open the secret elevator doors and set off a countdown! They had three minutes to get to safety!

Foxy sent everyone away, except Elise and Silva. They would not leave.

"Bitch, we ride together, we die together!" says Elise. Silva replies, "Is that a quote from *Bad Boys*." Elise says, "Well, we the Bad Girls." Sarcastically Foxy says, "REALLY! I already had a moment like this yesterday. So, stop selling crazy to me. I'm all stocked up here. We are talking about this shit right now, WOW!!!!!

Foxy showing a sadistic smile on her face and started trying to defuse the bomb that was ticking down on Chief Williams.

Silva says, "Just like old times, huh, girls!"

"Damn, we got to step off. Not enough time to defuse the bomb. That bitch knew that."

Elise and Silva looked at each other, then they both looked at Foxy. At the same time, all three said, "We need to run!"

They quickly freeze the C-4 with a canister that Foxy had on her person luckily.

Silva says, "Thank God for that shit you had in your pocket for mace."

By doing this, it gave them thirty to forty-five seconds longer to get out of harm's way.

Elise held on to Chief Williams and Silva carefully detached the straps with the booby trapped fuse attached to the bomb. When the liquid nitrogen falls below three degrees, it will explode. She carefully placed it into Foxy's hands. Foxy carefully placed it on the floor and they made a beeline to the elevator.

Then they quickly got on the elevator. They make it to the top. Now according to the set watch that was timed with the bomb, it shows 1 minute and 20 seconds and counting. They make it to the front door with 15 seconds. Down the steps, 5 seconds.

They jumped in the van with Rachel and Detective Skaggs puling the chief inside. Then they take off as the bomb goes off! Blowing up the building and all of Momma King's secret underground lair. It was laid to dust! As big fireballs came up from the ground.

Puffs of fireballs started coming from everywhere. As Rachel ducked and dodged fireballs here fireballs there; one fireball nipped the tail end of the van nearly getting them blown up! But by the skin of their teeth, they made it to safety. Finally, they make it on the other side of the property. Everyone is so happy that everybody made it out alive, as they get out of the van. There waiting on them was Mr. Remy and all of fifty-five men made it. They greeted them, then had a brief celebration. Then the police, fire trucks, and ambulances showed up moments later.

Then the director of the CIA shows up in a helicopter and lands near the police. He explained everything to the chief of police. Mr. Remy and his soldiers work for the CIA. Then goes on to explain that Det. Skaggs was not really Det. Skaggs! His name is Lieutenant Colonel Timothy Richardson of the USAF. Black Ops commander for the CIA. He was undercover on a mission to take down the Shadow Lady and the Pimp named Shawn Brown.

The police chief releases them. Chief Williams is rushed to the hospital at an undisclosed location. The girls said their goodbyes to the chief.

CHAPTER 38

The Return

Then Silva's phone rings. She answers saying, "Hello." The mystery person replies, "Guess who, sister from another mister." Silva says with great excitement, "Oh, baby girl, is it really you? Where are you right now?"

Kawanna nervously says, "At the airport. My sister isn't mad, is she?" Silva loudly replies in happiness, "Don't you move a muscle!" Silva quickly runs to Elise screaming with excitement.

Elise in complete surprise barely could understand what Silva was saying catching only one word her sister's name. Silva running at full speed toward Elise jumping up and down, up and down, screaming, "Kawanna, Kawanna, Kawanna!" Elise immediately grabs the phone, she screams out in such happiness saying, "Kawanna, where are you, baby sister?" Kawanna reluctantly replies, "I'm at the airport. Please do not bring Mom. I do not want her to fuss at me for ditching school, but I don't care. I'm in love, so please try to understand. Ok, sis." Not even thinking even caring about where she was or who she was with at that moment, she just wanted to hold her baby sister close to her heart. In a loving and gentle voice, Elise agrees saying,

"Ok, little sister, do not move. I am on my way, right now. Don't move, ok."

"Ok, ok. I will not move. Now hurry. I'm hungry," Kawanna replies. The ladies rush to the airport in the beat-up van.

Meanwhile, minutes later, a jet passes over the destroyed lair of Momma King. Sitting inside the jet was none other than the infamous Momma-King and her sidekick Shawn Brown, then her phone rings. There is a deep dark voice on the other end.

Momma King says, "How was your trip, Grand Master?"

The dark voice replies, "Overrated. My ETA to the meeting spot will be soon."

"Good, we have a lot to talk about," says Momma King. "Time to move to Phase two. Your protégé fought with great honor while sacrificing himself for the great mother Hera."

The dark voice replies, "He will be remembered amongst the dynasty of clans forever. Now I will finish where he could not."

Then Momma-King says, "I certainly hope so. Failure is not an option." Then the phone hangs up.

Meanwhile the ladies finally make it to the airport. There waiting by the baggage claim department was none other than Kawanna. They all ran to her and started hugging her and planting nonstop flurries of kisses all over her.

Elise says, "You had us so worried about you?"

"I took some time off to be with a very special friend," Kawanna replies.

Silva says, "A boyfriend?"

Kawanna replies, "Yes, for your information, Silva." Kawanna looks back at Elise, she says, "Before you fuss, I sent you my phone and a message, did you get it? I didn't want to tell you too much because … you know why, and I didn't tell the Mad Pack because I knew you would question them also, and I know they're CIA put to babysit me, but I didn't care because they made me feel important. Do not try to sit there and look surprised, sister."

Elise replies, "What? Aw ok, yeah. The CIA. You too much, little sister."

Then Kawanna says, "Ah-yeah, you made me that way, and I didn't want you and Mom to spook him. I took a well-deserved school break. Is Momma mad at me, sis? You told her, didn't you?"

Elise grabs Kawanna and holds her close, then replies, "Baby sister, we have a lot to talk about, sweetheart."

They all enter, lock their hands together, and walk toward the exit door of the airport. They abruptly run into a tall dark individual by accident. They quickly apologize to the stranger. He accepts their apology, and as the gentleman straightens up his jacket, the girls continue exiting the airport. The man grabs his bags, then walks out of the airport. For a moment, his sleeve moves back for a second revealing a mark, in likeness of the Red Scorpion Clan. Unlike the Red Scorpion, this tall dark figure had a mark of a white dragon. No one sees this event. Then he fades away into the chaotic streets of the city.